The Handbook of Environmental

Editors-in-Chief: O. Hutzinger · D. Barceló · A. Kostianoy

Volume 5 Water Pollution
Part S/1

The Handbook of Environmental Chemistry
Recently Published and Forthcoming Volumes

Environmental Specimen Banking
Volume Editors: S. A. Wise and P. P. R. Becker
Vol. 3/S, 2009

Polymers: Chances and Risks
Volume Editors: P. Eyerer, M. Weller
and C. Hübner
Vol. 3/V, 2009

The Black Sea Environment
Volume Editors: A. Kostianoy and A. Kosarev
Vol. 5/Q, 2008

Emerging Contaminants from Industrial and Municipal Waste
Removal Technologies
Volume Editors: D. Barceló and M. Petrovic
Vol. 5/S/2, 2008

Emerging Contaminants from Industrial and Municipal Waste
Occurrence, Analysis and Effects
Volume Editors: D. Barceló and M. Petrovic
Vol. 5/S/1, 2008

Fuel Oxygenates
Volume Editor: D. Barceló
Vol. 5/R, 2007

The Rhine
Volume Editor: T. P. Knepper
Vol. 5/L, 2006

Persistent Organic Pollutants in the Great Lakes
Volume Editor: R. A. Hites
Vol. 5/N, 2006

Antifouling Paint Biocides
Volume Editor: I. Konstantinou
Vol. 5/O, 2006

Estuaries
Volume Editor: P. J. Wangersky
Vol. 5/H, 2006

The Caspian Sea Environment
Volume Editors: A. Kostianoy and A. Kosarev
Vol. 5/P, 2005

Marine Organic Matter: Biomarkers, Isotopes and DNA
Volume Editor: J. K. Volkman
Vol. 2/N, 2005

Environmental Photochemistry Part II
Volume Editors: P. Boule, D. Bahnemann
and P. Robertson
Vol. 2/M, 2005

Air Quality in Airplane Cabins and Similar Enclosed Spaces
Volume Editor: M. B. Hocking
Vol. 4/H, 2005

Environmental Effects of Marine Finfish Aquaculture
Volume Editor: B. T. Hargrave
Vol. 5/M, 2005

The Mediterranean Sea
Volume Editor: A. Saliot
Vol. 5/K, 2005

Environmental Impact Assessment of Recycled Wastes on Surface and Ground Waters
Engineering Modeling and Sustainability
Volume Editor: T. A. Kassim
Vol. 5/F (3 Vols.), 2005

Oxidants and Antioxidant Defense Systems
Volume Editor: T. Grune
Vol. 2/O, 2005

Emerging Contaminants from Industrial and Municipal Waste

Occurrence, Analysis and Effects

Volume Editors: Damià Barceló · Mira Petrovic

With contributions by

M. L. de Alda · D. Barceló · J. Blasco · A. DelValls · M. Farre
M. Gros · M. Huerta-Fontela · M. Kuster · M. Petrovic · C. Postigo
J. Radjenovic · T. Smital · F. Ventura

Springer

Environmental chemistry is a rather young and interdisciplinary field of science. Its aim is a complete description of the environment and of transformations occurring on a local or global scale. Environmental chemistry also gives an account of the impact of man's activities on the natural environment by describing observed changes.
The Handbook of Environmental Chemistry provides the compilation of today's knowledge. Contributions are written by leading experts with practical experience in their fields. The Handbook will grow with the increase in our scientific understanding and should provide a valuable source not only for scientists, but also for environmental managers and decision-makers.
The Handbook of Environmental Chemistry is published in a series of five volumes:

Volume 1: The Natural Environment and the Biogeochemical Cycles
Volume 2: Reactions and Processes
Volume 3: Anthropogenic Compounds
Volume 4: Air Pollution
Volume 5: Water Pollution

The series Volume 1 The Natural Environment and the Biogeochemical Cycles describes the natural environment and gives an account of the global cycles for elements and classes of natural compounds.
The series Volume 2 Reactions and Processes is an account of physical transport, and chemical and biological transformations of chemicals in the environment.
The series Volume 3 Anthropogenic Compounds describes synthetic compounds, and compound classes as well as elements and naturally occurring chemical entities which are mobilized by man's activities.
The series Volume 4 Air Pollution and Volume 5 Water Pollution deal with the description of civilization's effects on the atmosphere and hydrosphere.
Within the individual series articles do not appear in a predetermined sequence. Instead, we invite contributors as our knowledge matures enough to warrant a handbook article.
Suggestions for new topics from the scientific community to members of the Advisory Board or to the Publisher are very welcome.

ISBN 978-3-642-09423-1 e-ISBN 978-3-540-74795-6
DOI 10.1007/978-3-540-74795-6

The Handbook of Environmental Chemistry ISSN 1433-6863

Softcover reprint of the hardcover 1st edition 2008

Cover design: WMXDesign GmbH, Heidelberg

Printed on acid-free paper

9 8 7 6 5 4 3 2 1 0

springer.com

Editors-in-Chief

Prof. em. Dr. Otto Hutzinger
Universität Bayreuth
c/o Bad Ischl Office
Grenzweg 22
5351 Aigen-Vogelhub, Austria
hutzinger-univ-bayreuth@aon.at

Prof. Andrey Kostianoy
P.P. Shirshov Institute of Oceanology
Russian Academy of Sciences
36, Nakhimovsky Pr.
117997 Moscow, Russia
kostianoy@mail.mipt.ru

Prof. Dr. Damià Barceló
Dept. of Environmental Chemistry
IIQAB - CSIC
Jordi Girona, 18–26
08034 Barcelona, Spain
dbcqam@iiqab.csic.es

Volume Editors

Prof. Dr. Damià Barceló
Dept. of Environmental Chemistry
IIQAB - CSIC
Jordi Girona, 18–26
08034 Barcelona, Spain
dbcqam@iiqab.csic.es

Mira Petrovic
Dept. of Environmental Chemistry
IIQAB - CSIC
Jordi Girona, 18–26
08034 Barcelona, Spain
mpeqam@cid.csic.es

Advisory Board

Prof. Dr. D. Barceló
Dept. of Environmental Chemistry
IIQAB - CSIC
Jordi Girona, 18–26
08034 Barcelona, Spain
dbcqam@iiqab.csic.es

Prof. Dr. P. Fabian
Lehrstuhl für Bioklimatologie
und Immissionsforschung
der Universität München
Hohenbachernstraße 22
85354 Freising-Weihenstephan, Germany

Dr. H. Fiedler
Scientific Affairs Office
UNEP Chemicals
11–13, chemin des Anémones
1219 Châteleine (GE), Switzerland
hfiedler@unep.ch

Prof. Dr. H. Frank
Lehrstuhl für Umwelttechnik
und Ökotoxikologie
Universität Bayreuth
Postfach 10 12 51
95440 Bayreuth, Germany

Prof. Dr. J. P. Giesy
Department of Zoology
Michigan State University
East Lansing, MI 48824-1115, USA
Jgiesy@aol.com

Prof. Dr. R. A. Hites
Indiana University
School of Public
and Environmental Affairs
Bloomington, IN 47405, USA
hitesr@indiana.edu

Prof. Dr. M. A. K. Khalil

Department of Physics
Portland State University
Science Building II, Room 410
P.O. Box 751
Portland, OR 97207-0751, USA
aslam@global.phy.pdx.edu

Prof. Dr. D. Mackay

Department of Chemical Engineering
and Applied Chemistry
University of Toronto
Toronto, ON, M5S 1A4, Canada

Prof. Dr. A. H. Neilson

Swedish Environmental Research Institute
P.O. Box 21060
10031 Stockholm, Sweden
ahsdair@ivl.se

Prof. Dr. J. Paasivirta

Department of Chemistry
University of Jyväskylä
Survontie 9
P.O. Box 35
40351 Jyväskylä, Finland

Prof. Dr. Dr. H. Parlar

Institut für Lebensmitteltechnologie
und Analytische Chemie
Technische Universität München
85350 Freising-Weihenstephan, Germany

Prof. Dr. S. H. Safe

Department of Veterinary
Physiology and Pharmacology
College of Veterinary Medicine
Texas A & M University
College Station, TX 77843-4466, USA
ssafe@cvm.tamu.edu

Prof. P. J. Wangersky

University of Victoria
Centre for Earth and Ocean Research
P.O. Box 1700
Victoria, BC, V8W 3P6, Canada
wangers@telus. net

The Handbook of Environmental Chemistry Also Available Electronically

For all customers who have a standing order to The Handbook of Environmental Chemistry, we offer the electronic version via SpringerLink free of charge. Please contact your librarian who can receive a password or free access to the full articles by registering at:

springerlink.com

If you do not have a subscription, you can still view the tables of contents of the volumes and the abstract of each article by going to the SpringerLink Homepage, clicking on "Browse by Online Libraries", then "Chemical Sciences", and finally choose The Handbook of Environmental Chemistry.

You will find information about the

- Editorial Board
- Aims and Scope
- Instructions for Authors
- Sample Contribution

at springer.com using the search function.

Color figures are published in full color within the electronic version on SpringerLink.

Preface

This book on "Emerging Contaminants from Industrial and Municipal Waste" is based on the scientific developments and results achieved within the European Union (EU)-funded project EMCO (reduction of environmental risks posed by emerging contaminants, through advanced treatment of municipal and industrial wastes). One of the key elements of the EMCO project was to provide support to the various Western Balkans countries involved in the project as regards the implementation of the Water Framework Directive (WFD) (2000/60/EC). A regional network, as proposed by the EMCO project, aiming to ensure the comparability (and reliability) of measurement data obtained by screening methodologies for water quality management, would support the EU Water Initiative, which aims to promote co-operation between countries in order to better manage their water resources.

The EMCO project addressed basically two directives: Directive 91/271/EEC to reduce the pollution in Community surface waters caused by municipal waste and the IPPC Directive (Directive 96/61/EC). This Directive expands the range of pollutants that should be monitored in industrial effluent discharges like those from the paper and pulp industry, refineries, textiles and many other sectors. The EMCO project has devoted its attention to the wastewater treatment technologies, especially in the Western Balkan countries. It is obvious that building up and improving wastewater treatment plant performance in the public and private sectors will avoid direct pollution of receiving waters by urban and industrial activities.

The book is divided into two volumes: Vol. I—Occurrence, Analysis and Effects, and Vol. II—Removal Technologies.

Volume I is structured in several chapters covering advanced chemical analytical methods, the occurrence of emerging contaminants in wastewaters, environmental toxicology and environmental risk assessment. Advanced monitoring analytical methods for emerging contaminants cover the use of liquid chromatography combined with tandem mass spectrometric detection or hybrid mass spectrometric techniques. It is certainly known that without these advanced mass spectrometric tools it would not be possible to investigate the fate and behaviour of emerging pollutants at the wastewater treatment plants and receiving waters at the nanogram per litre level. Ecotoxicology is also a very relevant aspect that should be taken into consideration for emerging contaminants, and it is also covered in this book. Risk assessment methodologies will allow us to critically establish the good performance of an appropriate wastewater treatment technology for the removal of urban, agricultural and industrial wastewaters.

Volume II covers different treatment options for the removal of emerging contaminants and includes membrane bioreactors (MBR), ozonization and photocatalysis, and advanced sorbent materials together with more conventional natural systems, such as artificial recharge and constructed wetlands. The MBR is an emerging technology based on the use of membranes in combination with traditional biological treatment. It is considered as a promising technology able to achieve more efficient removal of micro-pollutants in comparison to conventional wastewater treatment plants. Other examples reported in the book are advances in nanomaterials, also an emerging field in wastewater treatment, which are providing great opportunities in the development of more effective wastewater treatment technologies.

Overall, this book is certainly timely since the interest in emerging contaminants and wastewater treatment has been growing considerably during the last few years, related to the availability of novel treatment options together with the advanced and highly sensitive analytical techniques. This book can also be considered, in a way, the follow-up of two previous books in this series entitled *Emerging Organic Pollutants in Waste Waters and Sludge*, Vols. 1 and 2 (5 1 and 5 0), published in 2004 and 2005. The present book is complementary to these volumes since here much more attention has been devoted to wastewater treatment systems, which are a key part of this book.

The book will be of interest to a broad audience of analytical chemists, environmental chemists, water management operators and technologists working in the field of wastewater treatment, or newcomers who want to learn more about the topic. Finally, we would like to thank all the contributing authors of this book for their time and effort in preparing this comprehensive compilation of research papers.

Barcelona, September 2008

D. Barceló
M. Petrovic

Contents

Contents of Volume 5, Part S/2

Emerging Contaminants from Industrial and Municipal Waste

Removal Technologies

Volume Editors: Barceló, D., Petrovic, M.

ISBN: 978-3-540-79209-3

Hdb Env Chem Vol. 5, Part S/1 (2008): 1–35
DOI 10.1007/698_5_106

Published online: 18 April 2008

Emerging Contaminants in Waste Waters: Sources and Occurrence

Mira Petrovic[1,2] (✉) · Jelena Radjenovic[1] · Cristina Postigo[1] ·
Marina Kuster[1] · Marinella Farre[1] · Maria López de Alda[1] · Damià Barceló[1]

[1]Department of Environmental Chemistry, IIQAB-CSIC, c/ Jordi Girona 18–26,
08034 Barcelona, Spain
mpeqam@iiqab.csic.es

[2]Institució Catalana de Reserca i Estudis Avanzats (ICREA), Barcelona, Spain

Abstract There is a growing concern about possible ecotoxicological importance of various classes of emerging contaminants in the environment. Numerous field studies designed to provide basic scientific information related to the occurrence and potential transport of specific classes of emerging contaminants in the environment are being conducted with the aim to identify the sources and points of entry of these contaminants into the environment, and to determine their concentrations in both input streams (i.e., urban and industrial wastewaters) and receiving environment. This chapter summarizes the data regarding the occurrence of emerging contaminants in urban and industrial wastewaters, including some prominent classes such as pharmaceuticals, hormones, illicit drugs, surfactants and their degradation products, plasticizers, and perfluorinated compounds.

Keywords Emerging contaminants · Municipal waste waters · Occurrence · Sources

Abbreviations

AP	Alkylphenol
APEC	Alkylphenoxy carboxylates

APEO	Alkylphenol ethoxylate
BBP	Butylbenzyl phthalate
BE	Benzoylecgonine
BPA	Bisphenol A
CAFO	Concentrated animal feeding operation
CE	Cocaethylene
DA	Drug of abuse
DBP	Dibutyl phthalate
DEHP	Di(2-ethylhexyl) phthalate
DEP	Diethyl phthalate
DMP	Dimethyl phthalate
DnOP	Di-*n*-octyl phthalate
E1	Estrone
E2	Estradiol
E3	Estriol
EDC	Endocrine disrupting compound
EDDP	2-ethylidine-1,5-dimethyl-3,3-diphenylpyrrolidine perchlorate
EE2	Ethinylestradiol
FTOH	Fluorotelomer alcohol
LC-MS/MS	Liquid chromatography and tandem mass spectrometry
LSD	Lysergic acid diethylamide
MDE or MDEA	Methylenedioxyethylamphetamine
MDMA	3,4-Methylenedioxymetamphetamine hydrochloride
NP	Nonylphenol
NSAID	Non-steroidal anti-inflammatory drug
O-H-LSD	2-Oxo-3-hydroxy-LSD
OPEO	Octylphenol ethoxylate
OTC	Over-the-counter (drug)
PAEs	Phthalate acid ester
PEC	Predicted environmental concentration
PFBS	Perfluorobutane sulfonate
PFCA	Perfluoro carboxylic acid
PFCs	Perfluorinated compound
PFNA	Perfluorononanoic acid
PFOS	Perfluorooctane sulphonate
PhAC	Pharmaceutically active compound
POP	Persistent organic pollutant
THC	Δ^9-Tetrahydrocannabinol
WWTP	Wastewater treatment plant

1 Introduction

Until the beginning of the 1990s, non-polar hazardous compounds, i.e., persistent organic pollutants (POP) and heavy metals, were the focus of interest and awareness as priority pollutants and consequently were part of intensive monitoring programs. Today, these compounds are less relevant for the industrialized countries since a drastic reduction of emission has been achieved due to the adoption of appropriate measures and elimination of the dominant pollution sources.

However, the emission of so-called "emerging" or "new" unregulated contaminants has emerged as an environmental problem and there is a widespread consensus that this kind of contamination may require legislative intervention.

A wide range of man-made chemicals, designed for use in industry, agriculture, and as consumer goods and chemicals unintentionally formed or produced as by-products of industrial processes or combustion, are potentially of environmental concern. The term "emerging contaminants" does not necessarily correspond to "new substances", i.e., newly introduced chemicals and their degradation products/metabolites or by-products, but also refers to compounds with previously unrecognized adverse effects on the ecosystems, including naturally occurring compounds. Therefore, "emerging contaminants" can be defined as contaminants that are currently not included in routine monitoring programmes and which may be candidates for future regulation, depending on research on their (eco)toxicity, potential health effects, public perception and on monitoring data regarding their occurrence in the various environmental compartments [1].

Today, there are several groups of compounds that emerged as particularly relevant:

- Algal and cyanobacterial toxins
- Brominated flame retardants
- Disinfection by-products
- Gasoline additives
- Hormones and other endocrine disrupting compounds
- Organometallics
- Organophosphate flame retardants and plasticisers
- Perfluorinated compounds
- Pharmaceuticals and personal care products
- Polar pesticides and their degradation/transformation products
- Surfactants and their metabolites

For most emerging contaminants, occurrence, risk assessment, and ecotoxicological data are not available, and therefore it is difficult to predict what health effects they may have on humans and aquatic organisms. Numerous field studies designed to provide basic scientific information related to the occurrence and potential transport of specific classes of emerging contaminants in the environment are being conducted with the aim to identify the sources and points of entry of these contaminants into the environment, and to determine their concentrations in both input streams (i.e., urban and industrial wastewaters) and receiving environment.

The objective of this chapter is to give an overview of recent monitoring data, focusing on urban and industrial wastewaters. It reports the levels detected for some prominent classes such as pharmaceuticals, hormones, illicit drugs, surfactants and their degradation products, plasticizers and perfluorinated compounds. Possible sources and routes of entry of selected emerging contaminants into the environment are also discussed.

2 Pharmaceutical Residues

2.1 Sources

Pharmaceutically active compounds (PhACs) are an important group of emerging environmental contaminants that has been an issue of increasing interest in the international scientific community. In the European Union (EU), around 3000 different PhACs are used in human medicine (i.e., analgesics and anti-inflammatory drugs, β-blockers, lipid regulators, antibiotics, etc), thus their main route into the aquatic environment is ingestion following excretion and disposal via wastewater. After administration, pharmaceutical can be excreted as an unchanged parent compound, in the form of metabolites or as conjugates of glucuronic and sulphuric acid, primarily via urine and faeces. By analyzing the excretion pathways of 212 PhAC, equaling 1409 products, Lienert et al. [2] concluded that on average, 64% ($\pm$27%) of each PhAC was excreted via urine, and 35% ($\pm$26%) via faeces. In urine, 42% ($\pm$28%) of each PhAC was excreted as metabolites. Figure 1 shows the average total fraction excreted via urine and the fraction of the non-metabolized parent compound for selected therapeutic groups.

Metabolites of drugs can be expected to be bioactive and even more persistent, due to their increased polarity. Also, conjugates of parent compounds can be cleaved back into the original drug during the sewage treatment in wastewater treatment plants (WWTPs) [3]. Besides these WWTP discharges that are usually a consequence of their incomplete removal, other environmental exposure pathways of PhACs are manufacturing and hospital effluents, land applications (e.g., biosolids and water reuse), concentrated animal feeding operations (CAFOs), and direct disposal/introduction into the environment. For example, a survey conducted in the USA reported that the vast majority of people were disposing of expired medications via municipal garbage or domestic sewage [4].

In comparison to conventional priority pollutants, PhACs are designed to have specific pharmacologic and physiologic effects at low doses and thus are inherently potent, often with unintended outcomes in wildlife. They can undergo different chemical, photolytic, and biological reactions that mod-

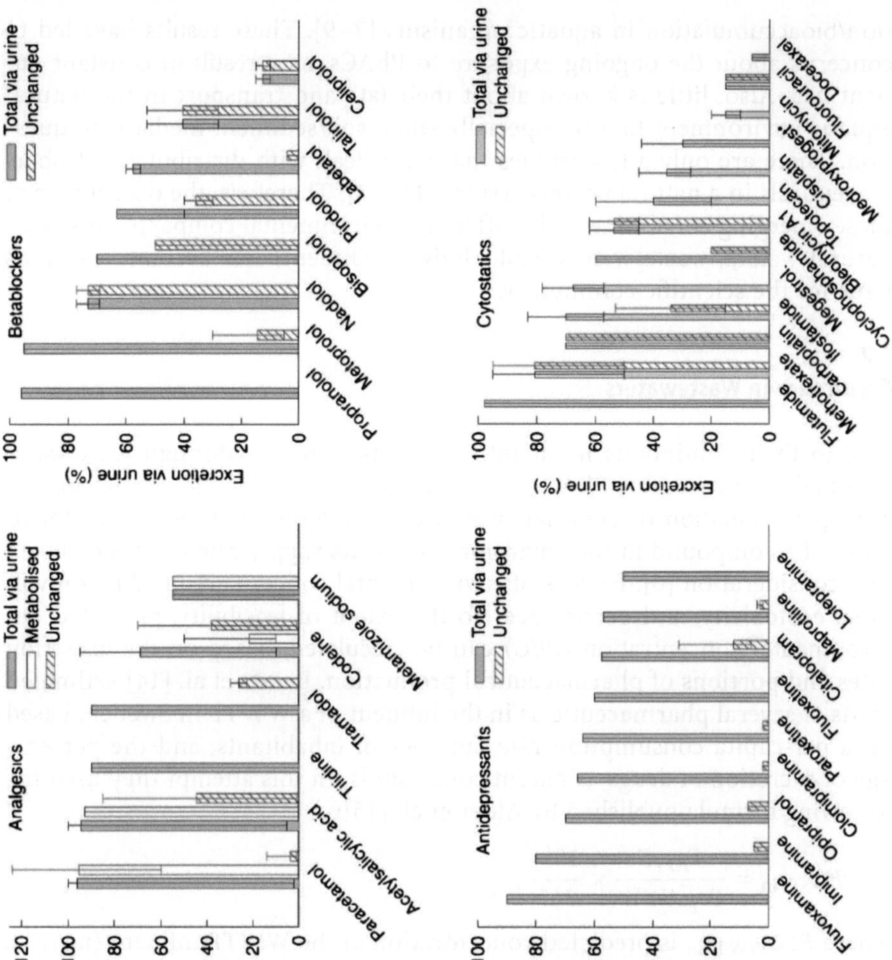

Fig. 1 Excretion via urine of selected therapeutic groups. The average for each PhAC is shown. *Error bars* denote the minimal and maximal value detected for each PhAC. The total fraction excreted via urine and the fraction of the non-metabolised parent compound (unchanged) is shown. For clarity, excretion via feces is not included. If bars are missing, then respective data were missing (e.g., no data on metabolism for the analgesic tilidine). For antidepressants, β-blockers, and cytostatics, metabolism data were missing for most PhAC. Cytostatics: cyclophosphamide includes cyclophosphane; *p*, medroxyprogesteronacetate. Reprinted with permission from [2]. © IWA Publishing 2007

ify the structure and physical transport of a compound in the environmental media. Furthermore, many PhACs do not exhibit acute toxicity but have a significant cumulative effect on the metabolism of non-target organisms [5] and ecosystem as a whole [6]. Some pharmaceuticals such as antidepressants, β-blockers or lipid regulators, can be prone to biococentration/bioaccumulation in aquatic organisms [7–9]. These results have led to concerns about the ongoing exposure to PhACs, as a result of constant patient use. Also, little is known about their fate and transport in the natural aquatic environment [5, 10], especially when soil/sediment media is in question. There are only a few studies that have dealt with distribution of pharmaceuticals in a natural porous system [11–13]. Therefore, the occurrence of these emerging contaminants in different environmental compartments (e.g., natural waters, waste waters, soil, sludge, sediment) has become a serious issue for the scientific community.

2.2
Occurrence in Wastewaters

Due to their continuous input into the aquatic media through wastewater as a main point-source, PhACs are considered to be "pseudo-persistent". In a proper evaluation of persistency of a certain compound, both transformation of a compound in the environment and its supply rate should be taken into consideration [6]. Factors of environmental concern are production volume, ecotoxicity, and persistence. To the extent of feasibility, predicted environmental concentration (*PEC*) can be calculated, based on the excretion rates and portions of pharmaceutical production. Bendz et al. [14] estimated loads of several pharmaceuticals in the influent of a WWTP in Sweden, based on a per-capita consumption rate, number of inhabitants, and the percentage of excretion of drugs as parent compounds. In this attempt they used the following formula published by Alder et al. [15]:

$$PEC_{\mathrm{STPin}} = \frac{F_{\mathrm{API}}E}{PopAWW} \times \frac{10^{12}}{365},$$

where PEC_{WWTPin} is predicted concentration in the WWTP influent ($\mathrm{ng\,L^{-1}}$), F_{API} consumption of β-blockers per year ($\mathrm{kg\,yr^{-1}}$), E fraction excreted as active substance without metabolization in urine and/or not absorbed (dimensionless), *Pop* population of Switzerland: 7.3 million inhabitants (cap) and *AWW* is amount of wastewater per capita and day ($400\,\mathrm{L\,cap^{-1}\,d^{-1}}$). The measured concentrations of some of them were of the same order of magnitude as the predicted ones (i.e., diclofenac, naproxen, and metoprolol). However, significantly lower concentrations of gemfibrozil, trimethoprim and atenolol, and significantly higher concentrations of carbamazepine were measured compared to the theoretical values. These discrepancies may be explained with seasonal variations in consumption rates and differences in excretion rates for humans depending on their age, sex, thyroid function, nutrition, etc [14]. In another study [16], predictions made out of excretion rates of atenolol (90%), sotalol (70%), metoprolol (5%) and propranolol (10%) and the data on their consumption in Switzerland gave PEC_{WWTPin} very similar to their measured concentrations in the influents of two Swiss WWTPs.

Estimations of pharmaceutical concentration in sewage have been usually performed by back-calculating the total prescribed mass from prescription rate data (number of defined daily doses) and excretion rates, partitioning, biodegradation, and the potential hydrolysis of conjugates [17, 18]. However, predictions based on annual sales of drugs are likely to be underestimating the loads of PhACs in the influents of WWTPs. This is because sales figures refer only to prescription drugs, and do not include over-the-counter drugs and Internet sales. Nevertheless, although these predictions have a high degree of uncertainty, they can focus attention on drugs that are candidates for further analytical studies.

The data on measured environmental loads of pharmaceutical residues is still scarce. The inputs of PhACs are generally considered to be constant and widely distributed. However, for some of them (e.g., antibiotics), differences between winter and summer influent loads were noted, probably because of higher attenuation in summer, and also less use of pharmaceuticals [19, 20]. On the other side, for other drugs (e.g., β-blockers, diuretics and anti-ulcer drugs) this seasonal variability was absent, which was consistent with the data on their occurrence [19].

Over the last 10 years, scattered data all over the world has demonstrated an increasing frequency of appearance in wastewater. The most ubiquitous drugs in WWTP influents are summarized in Table 1, together with their concentration ranges reported in literature.

The ubiquity of drugs is related to specific sales and practices in each country. For example, antihistamines, analgesics, and antidepressants are the families of drugs with major consumption in Spain, according to the National Health System. Indeed, in a study by Gros et al. [21] of the Ebro river basin, the highest influent loads from seven WWTPs were found for non-steroidal anti-inflammatory drugs (NSAIDs), lipid regulators, β-blockers and histamine H_1- and H_2-receptor antagonists. The total load of 29 monitored pharmaceuticals ranged from 1 to 5 g/day/1000 inhabitants for influent wastewater (Fig. 2). The results of a study in six WWTPs conducted in Italy [19] indicated high inputs of antibiotics sulfamethoxazole, ofloxacin, and ciprofloxacin, β-blocker atenolol, anti-histaminic ranitidine, diuretics furosemide and hydrochlorothiazide, and NSAID ibuprofen. A recent comprehensive reconnaissance of more than 70 individual wastewater contaminants in the region of Western Balkan (Bosnia and Herzegovina, Croatia, and Serbia) revealed the presence of 31 out of 44 analyzed pharmaceutical compounds at a concentration above the detection limit (typically 1–10 ng L^{-1}) [22]. The most abundant drug groups included analgesics and antiinflammatories, antimicrobials, β-blockers and lipid regulators, as shown in Fig. 3.

Generally, the most abundant loads are commonly reported for NSAIDs, which could be attributed to their wide consumption because they can be purchased without medical prescription (i.e., over-the counter (OTC) drugs).

Table 1 Occurrence of pharmaceutical residues in WWTP influents

Compound	Influent concentration ($\mu g\,L^{-1}$)	Refs.
Analgesics and anti-inflammatory drugs		
Ibuprofen	53.48–373.11; 150.73 [a]	[23]
	0.381–1.13; 0.672 [b]	[25]
	2.6–5.7	[134]
	8.45 [a]; 16.5 [c]	[38]
	23.4 [a]	[39]
	34–168; 84 [a]	[37]
Ketoprofen	0.108–0.369; 0.208 [b]	[25]
	0.146 [a]; 0.289 [c]	[38]
	2.9 [a]	[39]
	0.57 [c]	[40]
	0.16–0.97; 0.451 [a]	[28]
Naproxen	0.038–0.23; 0.1 [b]	[25]
	1.8–4.6	[134]
	8.6 [a]	[39]
	5.58 [a]; 17.1 [c]	[38]
Diclofenac	0.204 [a]; 1.01 [c]	[38]
	0.46 [a]	[39]
	3.25 [a]; 4.114 [a]; 3.19 [a]; 1.4 [a]; 0.905 [a]	[33]
	0.05–0.54; 0.25 [a]	[28]
	2.94 [c]	[40]
Indomethacin	0.23 [a]; 0.64 [c]	[38]
	nd	[28]
Acetyl-salicylic acid	0.47–19.4; 5.49 [b]	[25]
Salicylic acid	13.7 [a]; 27.8 [c]	[38]
Acetaminophen	0.13–26.09; 10.194 [a]	[28]
	29–246; 134 [a]	[37]
Lipid regulator and cholesterol lowering statin drugs		
Gemfibrozil	0.453 [a]; 0.965 [c]	[38]
	nd–0.36; 0.155 [a]	[28]
Bezafibrate	2.2 [a]	[39]
	1.96 [a]; 2.014 [a]; 6.84 [a]; 7.6 [a]; 1.55 [a]	[33]
	nd–0.05; 0.023 [a]	[28]
Clofibric acid	nd–0.11; 0.072 [a]	[28]
	0.36 [c]	[40]
Psychiatric drugs		
Carbamazepine	0.015–0.27; 0.054 [b]	[25]
	1.85 [a]; 1.2 [a]; 0.704 [a]; 0.67 [a]; 0.325 [a]	[33]
	nd–0.95; 0.42 [a]	[28]
	0.12–0.31; 0.15 [a]	[37]
Caffeine	52–192; 118 [a]	[37]

Table 1 (continued)

Compound	Influent concentration ($\mu g\,L^{-1}$)	Refs.
Antibiotics		
Sulfamethoxazole	nd–0.87; 0.59^a	[28]
Ofloxacin	nd	[28]
Ciprofloxacin	3.8^b; 4.6^c	[32]
Norfloxacin	0.17^b; 0.21^c	[32]
Trimethoprim	0.34^b; 0.93^c	[32]
	nd–4.22; 1.172^a	[28]
Antihistamines		
Ranitidine	nd–0.29; 0.188^a	[28]
β-blockers		
Atenolol	nd–0.74; 0.395^a	[28]
	$(0.971 \pm 0.03)^a$	[135]
Metoprolol	$(0.411 \pm 0.015)^a$	[135]
Sotalol	0.12–0.2; 0.167^a	[28]
	$(0.529 \pm 0.01)^a$	[135]
Propranolol	0.08–0.29; 0.168^a	[28]
	$(0.01 \pm 0.001)^a$	[135]
X-ray contrast media		
Iopromide	6.0–7.0	[134]
	$(7.5 \pm 1.5)^a$	[136]
Diatrizoate	$(3.3 \pm 0.7)^a$	[136]
Iopamidol	$(4.3 \pm 0.9)^a$	[136]

[a] mean,
[b] median,
[c] maximum concentrations.

For example, ibuprofen is usually detected at very high concentrations (in $\mu g\,L^{-1}$) [23–25]. Although the percentage of elimination of this drug is very high [21], it is still detected in rivers downstream WWTPs due to a very high usage in human medicine. Other very popular pain killers are acetaminophen (paracetamol) and aspirin (acetyl-salicylic acid). Acetyl-scalycilic acid is deacetylated in human organism into its more active form, salicylic acid, and two other metabolites, ortho-hydroxyhippuric acid and gentisic acid [26]. Ternes et al. [27] detected all three metabolites in sewage influent samples at very high $\mu g\,L^{-1}$ concentrations. Gros et al. [28] encountered an average concentration of 10.2 $\mu g\,L^{-1}$ in WWTP influents. The environmental loads of these drugs are expected to be substantially higher than the values predicted from their sales figures, as their use is often abused.

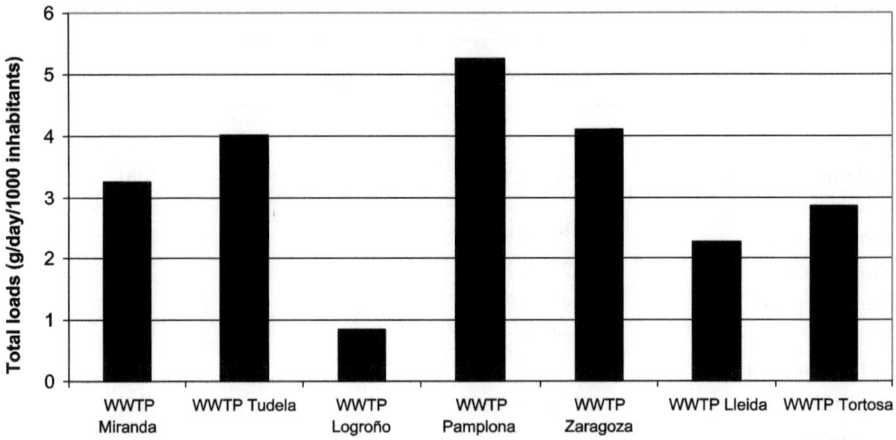

Fig. 2 Total loads of 29 multi-class pharmaceuticals, expressed as g/day/1000 inhabitants, measured in the raw wastewater entering seven major WWTP in the Ebro River basin. Modified from [21]

Besides these OTC drugs, pharmaceuticals ubiquitous in raw sewage are also prescription drugs β-blockers [21, 24, 29]. Atenolol seems to be the most frequently found β-blocker worldwide in WWTP influents [19, 30]. Atenolol, metoprolol, and propranolol were detected at high influent concentrations in a study by Nikolai et al. [30] (i.e., 110–1200, 170–520 and 20–92 ng L^{-1}, respectively). As far as their toxicity is concerned, it is suspected that mixtures of β-blockers are concentration-additive, since they all have the same mode of toxic action in the aquatic environment [31]. These drugs are also used in high quantities and are not efficiently eliminated in WWTPs, thus they are frequently encountered in surface waters [21].

Antibiotic losses to the environment are considered to be substantial due to their widespread consumption in human and veterinary medicine. Sulfamethoxazole, trimethoprim, ciprofloxacin, norfloxacin, and cephalexin had the highest median influent concentrations in a WWTP in Brisbane, Australia (360, 340, 3800, 170, and 4600 ng L^{-1}, respectively) [32]. Other studies confirmed high ubiquity of several antibiotics (i.e., ofloxacin, trimethoprim, roxyhtromycin and sulfamethoxazole) in sewage influent, though at low ng L^{-1} level [28, 33]. However, even at very low concentrations they can have significant ecotoxicological effects in the aquatic and terrestrial compartment [34, 35]. Indiscriminate or excessive use of antibiotics has been widely blamed for the appearance of so-called "super-bugs" that are antibiotic resistant. It is of crucial importance to control their emissions into the

Fig. 3 Frequency of detection for individual pharmaceuticals (%) in the Croatian wastewaters (modified from [22]) ▶

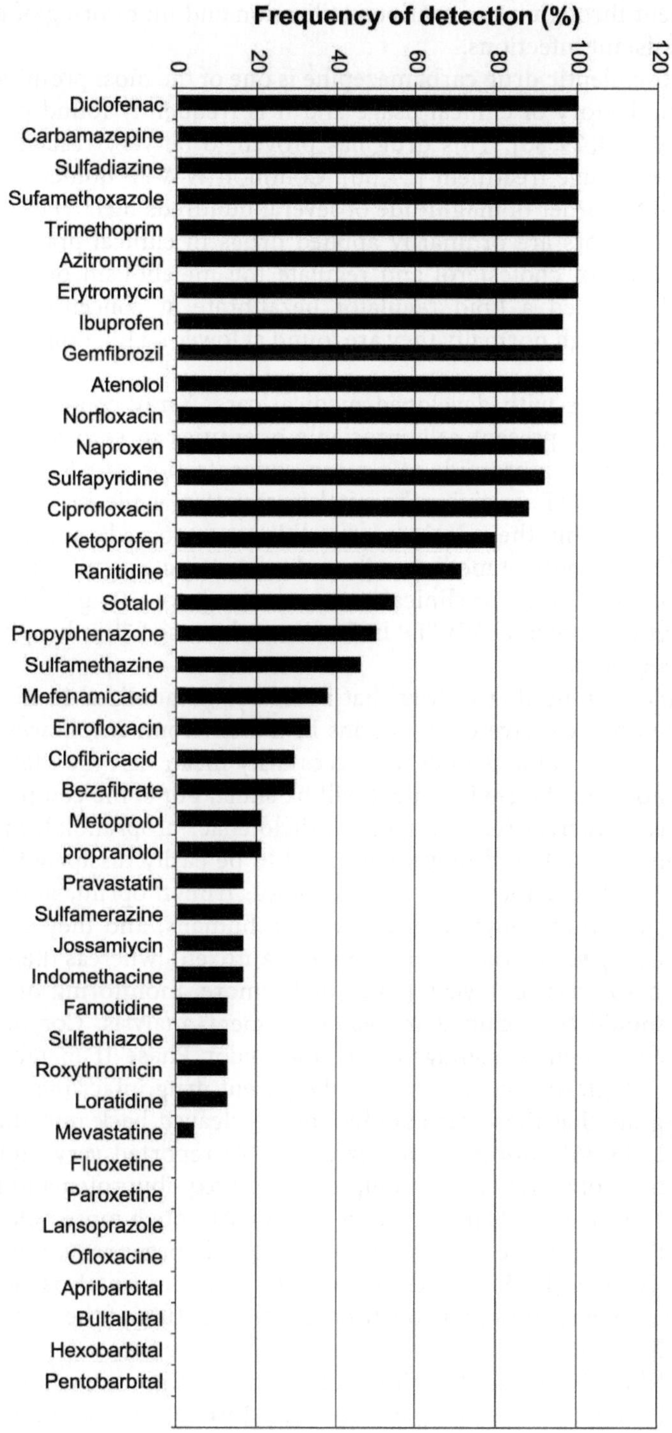
Frequency of detection (%)
0
20
40
60
80
100
120
Diclofenac
Carbamazepine
Sulfadiazine
Sufamethoxazole
Trimethoprim
Azitromycin
Erytromycin
Ibuprofen
Gemfibrozil
Atenolol
Norfloxacin
Naproxen
Sulfapyridine
Ciprofloxacin
Ketoprofen
Ranitidine
Sotalol
Propyphenazone
Sulfamethazine
Mefenamicacid
Enrofloxacin
Clofibricacid
Bezafibrate
Metoprolol
propranolol
Pravastatin
Sulfamerazine
Jossamiycin
Indomethacine
Famotidine
Sulfathiazole
Roxythromicin
Loratidine
Mevastatine
Fluoxetine
Paroxetine
Lansoprazole
Ofloxacine
Apribarbital
Bultalbital
Hexobarbital
Pentobarbital

environment through more cautious utilization and monitoring of outbreaks of drug-resistant infections.

The anti-epileptic drug carbamazepine is one of the most prominent drugs with a long history of clinical usage and it is frequently found in the environment [21, 24, 29, 36]. This drug has proven to be very recalcitrant since it by-passes sewage treatment [24, 36]. Common WWTP influent concentrations are in the order of magnitude of several hundreds ng L^{-1} [25, 28, 33, 37].

Lipid regulators are ordinarily applied drugs in clinical practice used to lower the level of cholesterol and regulate the metabolism of lipids. Clara et al. [33] detected a lipid regulator bezafibrate at concentrations up to 7.6 $\mu g\,L^{-1}$, although normally they are found at lower ng L^{-1} range [28, 33, 38–40].

In all countries with developed medical care, X-ray contrast media can be expected to be present at appreciable quantities in sewage water. Clara et al. [33] detected iopromide at a mean concentration of 3.84 $\mu g\,L^{-1}$ in the influent of a WWTP receiving hospital wastewater, while in WWTPs without a hospital within their drainage area this contrast media was not present. Iodinated X-ray contrast media are proved to contribute significantly total absorbable organic iodine in clinical wastewaters; up to 130 $\mu g\,L^{-1}$ of iodine in the influent of municipal WWTP in Berlin, and 10 mg L^{-1} in hospital sewage was detected [41].

We could assume that a drug that is highly metabolized in humans will be subjected to extensive degradations in the environment, however, a high metabolic rate in humans does not necessarily mean that the lifetime of the pharmaceutical in the environment will be short. For some compounds, this assumption is correct (e.g., ibuprofen, diclofenac, propranolol, metoprolol, and carbamazepine), and they were found to be easily dissipated in the environment [42]. On the other side, atenolol, trimethoprim, and naproxen are substances with a low metabolic rate in humans, and they are excreted mainly unchanged or as acyl-glucuronide (naproxen), whereas their half-lives range from 10 days to 1 year [43]. Furthermore, monitoring of metabolic products should be included in risk-assessment analysis. Commonly, glucuronide and sulphate conjugates are the major Phase II metabolites that leave the biologically active group of the parent drug intact [44]. Some evidence suggests that these metabolites can be cleaved back into the original compound [45, 46]. Moreover, Bendz et al. [14] reported very high influent concentrations of metabolites of ibuprofen, carboxy-ibuprofen and hydroxyl-ibuprofen (10.75 and 0.99 $\mu g\,L^{-1}$, respectively). Although more polar metabolites are presumed to be less hazardous to aquatic organisms, the European Medicines Agency (EMEA) guideline suggests environmental risk assessment of all human metabolites that constitute more than 10% of the total excretion of drug [47].

Due to their beneficial health effects and economic importance, the reduction of drug inputs into the environment through restricting or banning their use is not possible. Moreover, the use of pharmaceutical compounds is expected to grow with the increasing age of the population. The only possible way is to regulate their environmental pathways, perhaps on the source through labelling of medicinal products and/or developing disposal and awareness campaigns. Another option is to add sewage-treatment facilities in hospitals, and to enhance current wastewater-treatment techniques in order to eliminate more efficiently such polar pollutants.

3 Natural and Synthetic Estrogens

Estrogens are female steroid sex hormones based on a cholesterol structure. They are produced naturally in vertebrates in the gonads and adrenal cortex of both sexes and are responsible for the development of secondary sexual characteristics in the body. Their presence in the environment can cause negative effects to the endocrine functions of wildlife (e.g., aquatic organisms), posing an environmental risk. Estrogens reach the aquatic environment mainly due to incomplete removal in WWTP [48]. Other sources, such as livestock wastes will not be discussed in this section since these residues follow other pathways and do not end up in WWTPs.

3.1 Metabolism and Sources of Estrogens

In terms of binding to the human estrogen receptor, estradiol is the principal endogenous phenolic steroid estrogen. Estradiol is both metabolized reversibly and irreversibly. In the reversible metabolism, estradiol is transformed to estrone and estrone sulphate, meanwhile in the irreversible metabolism, estradiol is transformed to cathecol estrogens or estriol. These metabolites are mostly conjugated with glucuronides and, to a smaller extent, sulfates and excreted in the urine. A minor amount of the estrogens are excreted via feces as un-conjugated metabolites [49, 50].

Blocking the oxidation to estrone by, for instance, introducing an ethinyl group in position 17α or 17β of estradiol leads to much more stable products, which remain longer in the body. The consequence of this increased stability is that the so-formed synthetic steroid ethinylestradiol is excreted up to 80% unchanged in its conjugated form [51].

The human daily excretion of estradiol, estrone, and estriol vary from men (1.6, 3.9, 1.5 μg) to women (3.5, 8, 4.8 μg) maintaining similar proportions with estrone being the most abundant estrogen [5]. Pregnant women show a different profile with higher levels of estradiol and estrone by a factor of ten, and estriol daily excretion at 6000 μg. Women taking contraceptives based on ethinylestradiol excrete 35 μg of this synthetic estrogen daily [52].

In addition to the natural endogenous estrogens discussed above, other estrogens have to be taken into account, such as natural and/or synthetic estrogens administered as medicine. One of the main applications of estrogens is in contraceptives. The estrogen content in birth control pills is usually in the range of 20 to 50 μg daily [53]. Besides contraception, the uses of estrogens can largely be put into three main groups: the management of the menopausal and postmenopausal syndrome (its widest use); physiological replacement therapy in deficiency states; and the treatment of prostatic cancer in men and of breast cancer in postmenopausal women.

The main sources of estrogens to WWTPs are therefore from the natural production of estrogens by humans, from hormone and estrogen replacement therapies and the intake of hormone contraceptives containing ethinylestradiol.

3.2
Occurrence in Wastewater

The occurrence and environmental fate of estrogens have been reviewed in several articles [52, 54, 55]. The analysis of estrogens in wastewater has been discussed by Lopez de Alda et al. [56].

Estrogens are mainly excreted as their less active sulfate, glucuronide and sulfo-glucuronide conjugates [57]. However, in raw sewage and sewage-treatment plants (WWTPs), as well as in the environment, these conjugates may suffer deconjugation and act as precursors of the corresponding free steroids [58–61]. Thus, an appropriate evaluation of their occurrence and impact requires the analysis of both free and conjugated estrogens.

Most of the studies dealing with the investigation of estrogens in wastewaters have been performed in WWTPs receiving urban/domestic discharges and concentrations reported have been most usually in the ng/L range. Estradiol (E2) and estrone (E1) have been the free estrogens most frequently found, whereas estriol (E3) has been studied and detected only sporadically. However, E3 concentrations, when detected, have been usually higher than those of E2 and E1. In general, estrogens concentrations decrease in the order E3 > E1 > E2 (see Table 2 for examples). Thorough revision of all data available situates mean and median concentrations in the range of 9 to 20 ng/L for E2, 20 to 55 ng/L for E1 and 45 to 75 ng/L for E3 [58, 62–79].

The most studied synthetic estrogen, ethinylestradiol (EE2), has been either not detected [65, 67, 68] or detected at concentrations in general much lower than the other estrogens [58, 66, 77] (see Table 2). Levels higher than 100 ng/L have been only occasionally reported (e.g., 155 ng/L [63] and 138 ng/L [75]).

High levels of E1, E2, and E3 have also been reported by a few authors, e.g., 2100 ng/L of E2 [62], 200, 400, and 670 ng/L of E1 [62, 70, 79, 80] and 250 and 660 ng/L of E3 [79, 80].

Table 2 Levels of free estrogens in wastewater reported in some selected studies. Values are given as minimum–maximum (average or median) concentrations in ng L^{-1}

Estradiol	Estrone	Estriol	Ethinylestradiol	Refs.
3–22 (9)	8–52 (16)	n.a.	n.a.	[69]
10–31 (25)	16–60 (35)	23–48 (31)	n.d.	[68]
4.7–25 (12)	25–132 (52)	24–188 (80)	0.4–13 (3)	[58]
n.d.–21 (5.7)	10–57 (24)	27–220 (110)	n.d.	[67]
n.d.–234 (89)	9.4–232 (108)	n.d.–108 (23)	2.4–138 (57)	[75]

n.d. not detected;
n.a. not analysed

In general, it appears that the concentration of the un-conjugated estrogens in wastewater reflects roughly their excretion by the human body, where the high levels of estriol originate from pregnant women. This relation, however, is not found in influent wastewaters from WWTPs receiving industrial, or mainly industrial, wastes. In these cases, either estrone is the only estrogen detected [65] or the estrone concentration is significantly higher than that of estradiol and estriol [75].

The concentration of estrogens in wastewater entering WWTPs, together with other relevant data form the WWTP, such as influent flow-rate and the population served, has been used by some authors to calculate the loads of compounds (g/day) entering WWTPs. In a study dealing with the removal of pharmaceuticals, the calculated loads (mg/day/100 inhabitants) of estradiol (from not detected to 4), estrone (from not detected to 28) and ethinylestradiol (not detected) in six WWTPs were far below those of most of the other pharmaceuticals investigated [81]. Small loads of estrogens were also calculated by Ternes et al. [82] in a study performed in Germany (1 g/day E1, 0.5 g/day E2), and Brazil (5 g/day E1, 2.5 g/day E2).

In contrast to free estrogens, conjugated estrogen derivatives have been included only in a few studies [64, 65, 67, 74]. Mostly sulphates and glucuronides of E1, E2, and E3 have been included as target analytes and detected at similar levels as the free estrogens (see Table 3). Derivatives from the chemically more stable synthetic estrogen EE2 were studied by Gomes et al. [65], but no positive samples were found. Komori et al. [67] studied the presence of di-conjugated E2 derivatives and found high levels of the disulfate and moderately high levels of the sulfate-glucuronide derivative (see Table 3).

Although most estrogens are excreted as glucuronides the concentrations found at the entrance of WWTPs do not reflect this fact. Glucuronides levels are usually low; sulfates dominate the load of estrogens [74]. D'Azcenzo et al. [64] compared the amount of glucuronides and sulfates detected in female urine, a septic tank from a condominium and the entrance of a WWTP and found a higher percentage of sulfates (60%) at the entrance of the WWTP

Table 3 Levels ($ng\,L^{-1}$) of conjugated estrogen derivatives detected in waste water

Refs.	E1-3S	E2-S	E3-S	EE2-S	E1-G	E2-G	E2-2G	E3-G	EE2-G	E2-SG	E2-SS
[65]	10–14	n.a.	n.a.	n.d.	n.d.	n.a.	n.a.	n.d.	n.d.	n.a.	n.a.
[74]	34	3.2	n.a.	n.a.	0.4	0.3	n.a.	n.a.	n.a.	n.a.	n.a.
[64]	27	9	47	n.a.	10	n.d.	9	39	n.a.	n.a.	n.a.
[67]	42	110	22	n.a.	11	18	n.a.	22	n.a.	5.5	77

S, sulphate;
G, glucuronide;
n.a., not analysed;
n.d., not detected

than in the septic tank (55%) and the female urine (22%), suggesting that glucuronides might be de-conjugated in the sewer moiety and reach the WWTP at lower levels. In contrast, sulfates appear to be more stable than glucuronides, probably because bacterial sulfatases are present at lower concentrations than glucuronidases and/or because they have low affinity towards steroid sulfates. One example presented by Huang et al. [83] showed that sulfatases enzymes convert only 30% of E2 sulfate into E2.

In conclusion, the levels of estrogens in wastewater are occasionally very high (>100 ng/L), although in average values are usually below 100 ng/L. The calculated loads of estrogens entering the WWTPs are relatively low compared to those of pharmaceutical residues. However, there is no sufficient data on the concentration of the conjugated derivatives and their loads. Their de-conjugation can pose a problem if elimination is not complete.

4 Drugs of Abuse

According to the World Drug Report 2007, about 200 million people use illicit drugs each year globally. Drugs of abuse (DAs) consumption seems now to be stabilized after the increasing trends observed over a decade [84, 85]. Similar to PhACs, these substances are considered to be "pseudo-persistent" in the environment, thus they have become a group of emerging environmental contaminants of interest. DAs reach aquatic systems mainly through sewage water. After drug ingestion, diverse proportions of the parent compound, conjugated forms and metabolites are excreted via urine and flushed towards municipal WWTPs. Some of them may not be efficiently or completely removed at WWTPs and therefore they will be released into the environment via WWTP effluents. In addition to WWTPs discharges, direct disposal into the environment is to a lesser extent another pathway to the aquatic media.

The toxicological or cumulative effect of these substances on the ecosystem has not yet been studied. These compounds have specific physiologic and psychological effects in humans at low-concentration doses (mg or even μg in the case of lysergic acid diethylamide), thus the evaluation of the exposure of the wildlife to the bioactive molecules may be of interest, according to their occurrence in the environment. Fate and transport in aquatic environments is also not known. Most of them are polar compounds that will be concentrated in aqueous environmental matrices; however, some of them, such as cannabinoids, are likely to bioaccumulate in organisms or concentrate in sediments due to their physico-chemical properties (octanol–water partition coefficient, solubility...). A study of the distribution of these compounds in the different environmental compartments may also be a matter of scientific interest.

Since 2004, several authors have developed analytical methodologies based on liquid chromatography and tandem mass spectrometry (LC-MS/MS) detection to evaluate the occurrence of drugs of abuse in sewage and natural waters [86–92]. The target drugs of abuse and metabolites studied so far belong to five different classes: cocainics, amphetamine-like compounds, opiates, cannabinoids, and lysergics. Although a lack of data on drugs of abuse residues in environmental waters is still remarkable, mean values of these substances reported so far in the peer-reviewed literature are summarized in Table 1. The table gathers levels of common drugs of abuse and their metabolites detected in influent waters collected at different European WWTPs located in Spain [86, 92], Ireland [88], Italy [87, 89], Switzerland [87] and Germany [90].

The ubiquity of the different target compounds is directly related to local patterns of drug abuse. The highest loads, thus the highest consumption, are usually reported for two cocainic compounds, namely, cocaine and its main metabolite benzoylecgonine (BE), that are commonly detected at the high $ng\,L^{-1}$ or even the $\mu g\,L^{-1}$ level. The highest concentrations have been found in influent waters collected at a WWTP located in Barcelona, where BE, an inactive metabolite of cocaine with a relatively long half-life, was present at a mean concentration of $4226\,ng\,L^{-1}$ [92]. Cocaethylene (CE), which is a transesterification product of cocaine formed when cocaine is consumed together with ethanol, has not been detected at high levels; thus either this practice is rather limited or, what is more likely, CE transforms rapidly into metabolites not studied yet in WWTPs, such as norcocaethylene and ecgonine ethyl ester. Other cocaine metabolites, norcocaine and norbenzoylecgonine, have been studied at two WWTPs in Italy but their levels did not surpass $40\,ng\,L^{-1}$.

From the studied opiates, only morphine has been found in some WWTPs at high $ng\,L^{-1}$ levels, resulting probably from its medical applications. Although morphine is excreted in urine mainly as glucuronide metabolites, cleavage of the conjugated molecules in wastewater is likely to occur in the light of the low levels found for morphine-3β-d-glucuronide (the only conjugated compound studied) in comparison with those usually detected for morphine [87]. Heroine has been either not detected or detected at very low concentrations due to its low consumption and its also rapid hydrolysis to morphine and 6-acetylmorphine (heroine is quite unstable in blood serum) [93]. The results of the study done in WWTPs located in Italy and Switzerland [87] indicate that methadone, that is a long-acting opioid agonist used for treating acute and chronic pain and for preventing opiate withdrawal, is commonly present at lower levels than its pharmacologic inactive metabolite 2-ethylidine-1,5-dimethyl-3,3-diphenylpyrrolidine perchlorate (EDDP); both compounds were found in both areas at ng L^{-1}.

Concerning lysergic acid diethylamide (LSD) and its metabolites nor-LSD and nor-iso LSD (nor-LSD) and 2-oxo-3-hydroxy-LSD (O-H-LSD), absence or very low concentrations have been reported in influent samples. These results are in line with the very low doses of LSD needed to produce an effect compared to those needed in the case of other drugs (μg vs. mg), as LSD is the most potent psychoactive drug known so far [93].

The most abundant amphetamine-like compound detected in influent sewage waters is the phenylethylamine ephedrine. Besides a recreational and illicit use, this drug presents medical applications as topical decongestant and bronchodilator in the treatment of asthma and in the reversal of hypotension states. The so-called "designer drugs" 3,4-methylenedioxymetamphetamine hydrochloride (MDMA or "ecstasy"), methylenedioxyethylamphetamine (MDE, MDEA or "Eve") and 3,4-methylenedioxyamphetamine (MDA or "Love pills", and metabolite of both MDE and MDMA), have been detected frequently at the ng L^{-1} level in the different studied WWTPs. As shown in Table 4, amphetamine and methamphetamine are usually present in this type of matrix at lower concentrations than MDMA.

The presence of Δ^9-tetrahydrocannabinol (THC), which is the most psychologically active constituent of Cannabis (the most widely used illicit drug), in influent sewage waters has been observed insignificant as compared to that of its metabolites since THC is extensively metabolized before excretion. 11-nor-9 carboxy THC (nor-THC) is the major THC urinary metabolite and 11-hydroxy-THC (OH-THC) is the main psychoactive metabolite in the body. Thus, monitoring of these metabolites seems to be more appropriate to study the occurrence of cannabinoids in waters.

Measured values of DAs in sewage waters provide real-time data to estimate drug abuse at the community level. This strategy was first proposed by Daughton in 2001 [94] and implemented 4 years later by Zucatto et al. [89] to estimate cocaine abuse in the north of Italy. Such estimations, obtained in a fairly cheap and anonymous way (avoiding potential privacy conflicts), allow the immediate adoption of appropriate measures by the responsible authorities to fight drug abuse by the population. Efficiency of removal of DAs in WWTPs is largely unknown and should be addressed in order to control their

Table 4 Occurrence of drugs of abuse residues in WWTPs influents

Compound	Concentration ($ng\,L^{-1}$)	Refs.
Cocainics		
Cocaine	225^{a}, 79^{b}	[86]
	$(421.4\pm83.3)^{b}$, $(218.4\pm58.4)^{b}$	[87]
	$(489\pm117)^{b}$	[88]
	42–120; 80.25^{b}	[89]
	$(860.9\pm213.6)^{b}$; 502.3^{b}	[92]
Norcocaine	$(13.7\pm5.3)^{b}$; $(4.3\pm0.9)^{b}$	[87]
Benzoylecgonine	2307^{a}, 810^{b}	[86]
	$(1132.1\pm197.2)^{b}$, $(547.4\pm169.4)^{b}$	[87]
	$(290\pm11)^{b}$	[88]
	390–750; 550^{b}	[89]
	78	[90]
	$(4225.7\pm1142.8)^{b}$; 1456.7^{b}	[92]
Norbenzoylecgonine	$(36.6\pm7.8)^{b}$, $(18.8\pm5.6)^{b}$	[87]
Cocaethylene	$(11.5\pm5.1)^{b}$, $(5.9\pm2.6)^{b}$	[87]
	$(77.5\pm33.2)^{b}$, $(78.5)^{b}$	[92]
	n.d.	[88]
Opiates		
Heroine	n.d., 2.4^{b}	[92]
Morphine	$(83.3\pm11.8)^{b}$, $(204.4\pm49.9)^{b}$	[87]
	n.d	[88]
	820^{a}; 310^{c}	[90]
	$(162.9\pm20.0)^{b}$, 68.1^{b}	[92]
6 Acetyl morphine	$(11.8\pm8.5)^{b}$, $(10.4\pm4.8)^{b}$	[87]
	$(12.8\pm3.1)^{b}$, 8.4^{b}	[92]
Morphine-3β-d-glucuronide	$(2.5\pm7.1)^{b}$, $(18.1\pm30)^{b}$	[87]
Methadone	$(11.6\pm1.7)^{b}$, $(49.7\pm9.6)^{b}$	[87]
	n.d.	[88]
EDDP	$(19.8\pm3.1)^{b}$, $(91.3\pm19.2)^{b}$	[87]
	n.d.	[88]
Amphetamine-like compounds		
Amphetamine	15^{a}; 15^{b}	[86]
	$(14.7\pm10.6)^{b}$; <LOQ	[87]
	$(41.1\pm9.1)^{b}$; 20.8^{b}	[92]
Methamphetamine	$(16.2\pm7.1)^{b}$; <LOQ	[87]
	$(18.2\pm5.8)^{b}$; 4.8^{b}	[92]
	n.d.	[86]
MDMA	91^{a}; 49^{b}	[86]
	$(14.2\pm14.5)^{b}$, $(13.6\pm12.6)^{b}$	[87]
	$(133.6\pm29.8)^{b}$, $(135.13)^{b}$	[92]
	n.d.	[88]

Table 4 (continued)

Compound	Concentration ($ng\,L^{-1}$)	Refs.
MDEA	27^{a}; 28^{b}	[86]
	$(1.5 \pm 3.8)^{b}$, < LOQ	[87]
MDA	$(4.6 \pm 7.3)^{b}$, < LOQ	[87]
Ephedrine	$(591.9 \pm 124.5)^{b}$, 399.3^{b}	[92]
LSD and its metabolites		
LSD	$(2.8 \pm 1.2)^{b}$, 2.9^{b}	[92]
	n.d.	[86]
	n.d.	[88]
2-oxo-3-hydroxy-LSD	$(5.6 \pm 12.1)^{b}$, 3.4^{b}	[92]
Nor-LSD & nor-iso LSD	$(4.3 \pm 1.8)^{b}$, 13.5^{b}	[92]
Cannabinoids		
THC	nd; 14.24^{b}	[92]
11-nor-9-carboxy-THC	$(62.7 \pm 5)^{b}$; $(91.2 \pm 24.7)^{b}$	[87]
	$(4.3 \pm 7.8)^{b}$; 21.03^{b}	[92]
11-hydroxy-THC	$(8.4 \pm 2.1)^{b}$; 46.3^{b}	[92]

[a] maximum concentration,
[b] mean,
[c] median

release to the environment and avoid potential adverse effects in the aquatic ecosystem.

5
Surfactants (Alkylphenol Ethoxylates and Related Compounds)

Surfactants are produced in huge amounts and used in households as well as in industrial cleansing processes and as such they make up one of the most relevant organic pollutants of anthropogenic origin with the high potential to enter the environment. After use, detergents are usually discarded down the drain into sewer systems and afterwards treated in WWTP where they are completely or partially removed by a combination of sorption and biodegradation.

Among various classes of non-ionic, anionic, and cationic surfactants, alkylphenol ethoxylates (APEOs) are the group that raised the most concern. APEOs are effective nonionic surfactants, widely used as industrial cleaning agents and wherever their interfacial effects of detergency, (de)foaming, (de)emulsification, dispersion or solubilization can enhance products or process performance. Although parent APEOs are not classified as highly toxic substances (EC_{50}, 48 h, *Daphnia magna* 1.5 mg L^{-1}) their environmental acceptability is strongly disputed because of estrogenic metabolic products (alkylphenols (APs) and carboxylic derivatives (APECs)) generated during wastewater treatment. Because of these findings, APEOs are banned or restricted in Europe. Throughout northern Europe (Scandinavia, UK, and Germany) a voluntary ban on APEO use in household cleaning products began in 1995 and restrictions on industrial cleaning applications in 2000 [95]. This resulted in a significant reduction of APEO concentrations found. For example, in five Norwegian WWTP nonylphenol (NP) was found in the range of 0.2–7 μg L^{-1} in the effluent samples in 2002, while concentrations below the detection limit (2 ng L^{-1}) were found in the 2004 samples [96], which is attributed to new restrictions implemented in 2002. Similarly, the NP concentrations in digested sewage sludge in Switzerland were around 1.3 g/kg dry sludge before the ban of NP surfactants in laundry detergents in 1986. In the 1990s, the NP concentrations in sludge ranged from 0.1 to 0.2 mg/kg dry sludge [97]. In Catalonia (Spain), typical levels of NP measured in WWTPs in 1998 and 1999 ranged from 100 to 200 μg L^{-1} in influents, while 2002–2003 data show almost a 10-fold decrease (Fig. 4), which suggests a gradual withdraw and replacement of NPEOs by Spanish tanneries and textile industry [98].

However, mainly because of lower production costs, APEOs are still being used in substantial amounts in institutional and industrial applications. Hence information about the total concentrations of APEOs and their degradation products in environmental matrices is essential in assessing the environmental impact of these compounds.

Several extensive monitoring programs were conducted with the objective of determining the concentrations of APEO and their degradation products in raw and treated wastewaters. The concentrations of NPEOs (Table 5) in WWTP influents varies from less than 30 to 1035 μg L^{-1}. In industrial wastewaters (especially from tannery, textile, pulp, and paper industry) much higher values, up to 22 500 μg L^{-1}, are detected. Octylphenol ethoxylates (OPEOs) typically comprised 5–15% of total APEOs in WWTP influents, which is congruent with their lower commercial use. Concentrations found in WWTP effluents rarely exceeded 100 μg L^{-1}, corresponding to an elimination of the parent compound ranging from 80–98%.

However, their removal led to the formation of transformation products that are much more resistant to further microbial degradation. Acidic and neutral degradation products of NPEOs have been found to be rather resistant to further degradation, being NP the most recalcitrant intermediate. NPEO metabolites, NP and NPECs are already detected in WWTP influents, due to in-sewer degradation, in concentrations up to 40 μg L^{-1}. Recently, a comprehensive study in the region of Western Balkan (Bosnia and Herzegovina, Croatia, and Serbia) [22] showed widespread occurrence of surfactant-derived alkylphenolic compounds, although the concentration levels were

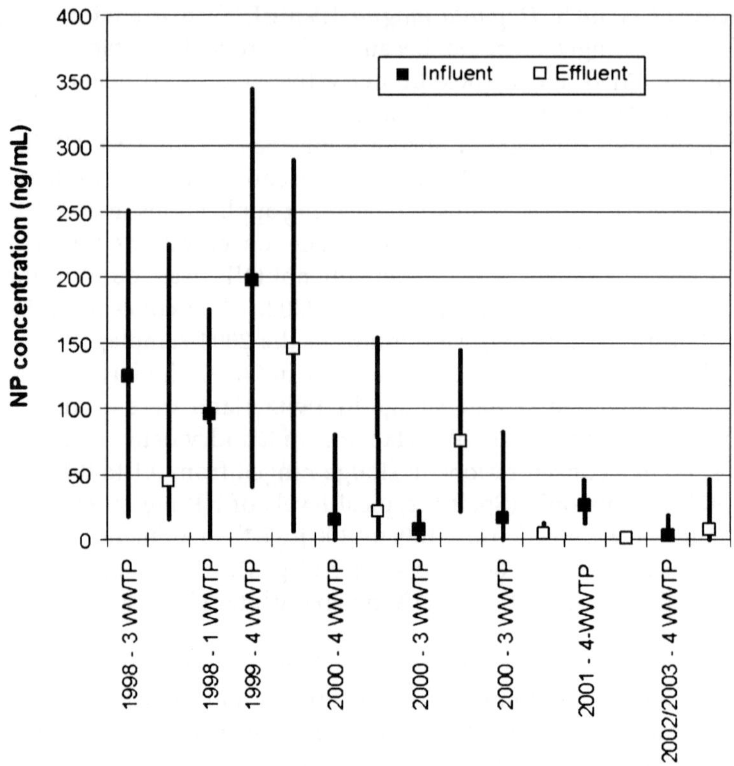

Fig. 4 Concentration of NP in influents and effluents of WWTP in Catalonia (NE Spain) in the period from 1998 to 2003 (Adapted from [98])

relatively low and suggest a decreasing trend in comparison to some previous campaigns conducted in early 1990s [99]. The concentration of NP, as the most toxic and most potent estrogen disrupting compound derived from NPEO surfactants [100], was present in concentrations up to 4.4 $\mu g L^{-1}$ with an average value of 1.7 $\mu g L^{-1}$. It is interesting to mention that Croatia was one of the first countries that introduced water-quality criteria for NP with a maximum permissible concentration in ambient water of 1 $\mu g L^{-1}$ [101], 15 years before it was accepted as a priority pollutant in the EU Water Framework Directive. Besides NP, all municipal wastewaters contained measurable levels of other metabolites derived from NPEO surfactants, in particular NPEC. The composition of alkylphenolic compounds was highly variable and revealed a strong impact of various biotransformation and physico-chemical processes on the distribution of individual alkylphenolic compounds in various types of wastewater samples. The most abundant alkylphenolic species in non-treated wastewaters was NP, while NPEC were the dominant species in biologically treated effluents, which is in agreement with earlier reports on this subject [102].

Table 5 Concentration ranges of alkylphenolic surfactants and their metabolites in raw wastewater entering WWTP

Compounds	Country	Concentration ($\mu g\, L^{-1}$)	Refs.
NPEO	Germany	120–270	[137]
	Austria	2.6–35 (NP_1EO)	[138]
		1.2–5.8 (NP_2EO)	
	Italy	29–145	[139]
		127–221	[140]
	Spain	27–880 (2120) [a]	[141–143]
	Switzerland	96–430	[144, 145]
	The Netherlands	< 0.1–125	[146]
		50–22500 [a]	
	Croatia	5–392	[22]
NPEC	Spain	< 0.2–14 [a]	[147]
		< 0.4–219	[141, 143]
	Croatia	< 0.001–3.20 (NPE_1C)	[22]
		< 0.001–4.37 (NPE_2C)	
NP	Belgium	< 0.4–219 [b]	[148]
	Italy	2–40	[149]
	Spain	< 0.5–22	[141, 147]
		17–251 [a]	[143]
	The Netherlands	< lod-19 (40) [a]	[146]
	Croatia	0.460–4.40	[22]
	Norway	< 0.002–5.2	[96]
	Austria	1.05–8.6	[138]

[a] WWTP receiving high percentage of industrial wastewater
[b] effluent of a textile plant

Octylphenolic analogues of NPEOs and their metabolites represented only a small percentage of the total alkylphenolic compounds in all analyzed samples, typically less than 10%. This is important for the assessment of the endocrine disrupting potential associated with APEO surfactants and their metabolites, because OP is an endocrine disrupting compound (EDC) four times more potent than NP [100].

6 Perfluorinated Compounds

Perfluorinated compounds (PFCs) have been manufactured for more than 50 years, and released into the environment following production and use. As a result, PFCs are now acknowledged to be widespread environmental contaminants. PFCs repel both water and oil and these compounds are therefore ideal chemicals for surface treatments. These compounds have been used for many industrial applications such as stain repellents (such as Teflon), textile, paints, waxes, polishes, electronics, adhesives, and food packaging.

PFCs are both hydrophobic and lipophobic, and are highly stable in the environment. Many of the degradation products of PFCs have been found in the environment throughout the world, because of the strong carbon–fluorine (C–F) bond associated with FASs. In addition, the most important PFC: perfluorooctane sulphonate (PFOS) and perfluoro carboxylic acids (PFCAs) are also stable degradation products/metabolites of neutral PFC. These precursor compounds are more volatile and therefore more likely to undergo long-range atmospheric transport, with sufficient atmospheric lifetimes to reach remote locations, where they can break down.

Possible precursor compounds for PFCAs and PFOS are fluorotelomer alcohols (FTOHs). Fluorotelomeralcohols are manufactured as a raw material used in the synthesis of fluorotelomer-based surfactants and polymeric products. The manufacture of FTOHs usually results in a mixture containing six to 12 fluorinated carbon congeners, the 8 : 2 FTOH being the dominant one. Release of the volatile FTOH may occur all along the supply chain from production application.

PFOS and PFOA are environmentally persistent substances that have been detected worldwide in human blood, water, soils, sediments, air, and biota samples [103].

PFCs are currently receiving great attention because of their persistence [104, 105]], bioaccumulation [106], and potential health concerns including toxicity [107] and cancer promotion [108], and they are now included in different health programs in EEUU to provide a better assessment of the distribution, toxicity, and persistence of these compounds in humans [109]. Research questions include understanding the sources of perfluorinated compounds and their environmental fate and transport.

In the EU, there is currently no legislation on the use of PFCs associated with their (potential) environmental and/or human health effects. It should, however, be noted that some legislation which generally applies to the release of substances to the environment may be relevant to the release of PFOS. This is the case with the IPPC Directive 96/61/EC concerning integrated pollution prevention and control, which includes fluorine and its compounds in the "indicative list of the main polluting substances to be taken into account if they are relevant for fixing emission limit values" (Annex III to the Directive) [110].

Recent studies have attempted to explain the occurrence of PFOA in the Arctic environment by oceanic transport as a result of the manufacture and use of PFOA [104, 111, 112]. Armitage et al. assumed emissions via waste water treatment plants effluents and their predictions have indicated PFOA concentrations in the Northern Polar Zone (equivalent to the Arctic Ocean) would increase until about 2030 and then gradually decline as ocean concentrations adjust to lower emission rates.

Table 6 Concentrations (ng L^{-1}) of perfluorinated compounds found in wastewaters and different environmental waters

Type of water	Country and site	PFOS	PFOA	PFHpA	PFNA	PFDA	Refs.
Wastewater							
Effluent	Austria	4.5–20	10–21	2.5–4.6	0–2	0–2	[150]
Effluent	EEUU (New York)	3–68	58–1050		0–376	0–47	[114]
Effluent	EEUU (Kentucky)	8–993	8.3–334	–	0–15.7	0–201	[115]
Effluent	EEUU (Georgia)	0–70	7–227	–	0–54	0–86	[115]
River							
Dalälven	Sweden	–	< 0.97	0.36	< 0.14	–	[151]
Vindelälven	Sweden	–	< 0.65	0.2	0.22	–	[151]
Elbe	Germany	–	7.6	2.7	0.27	–	[151]
Oder	Poland	–	3.8	0.73	0.73	–	[151]
Vistula	Poland	–	3.0	0.48	0.36	–	[151]
Po	Italy	–	200	6.6	1.46	–	[151]
Danuve	Romania/ Ucrania	–	16.4	0.95	0.27	–	[151]
Daugava	Letonia	–	< 2.2	0.86	0.36	–	[151]
Seine	France	–	8.9	3.7	1.26	–	[151]
Loire	France	–	3.4	0.90	0.43	–	[151]
Thames	UK	–	23	4.1	0.79	–	[151]
Rhine	Germany	–	12.3	3.3	1.50	–	[151]
Guadalquivir	Spain	–	4.6	1.58	1.02	–	[151]
Rhine	Germany (Breisach)	26	2	–	–	–	[120]
Rhine	Germany (Mainz)	12	3	–	–	–	[120]
Rhine	Germany (Ludwigshafen)	5	2	–	–	–	[120]
Ruhr	Germany (Duisburg)	5	48	–	–	–	[120]
Ruhr	Germany (Schwerte)	14	177	–	–	–	[120]
Elpe	Germany (Bestwig)	–	1168	–	–	–	[120]
Moehne	Germany (Heidelberg)	193	3640	148	–	–	[120]
Tenjin	Japan	4.7	39	–	–	–	[152]
Katsura	Japan	< 5.2	7.9	–	–	–	[152]
Lake							
Shihwa	Korea	89.11	19.22	2.50	3.26	1.98	[153]
Maggiore	Italy	7.8	2.4	2.4	0.6	3.7	[153]

Table 6 (continued)

Type of water	Country and site	PFOS	PFOA	PFHpA	PFNA	PFDA	Refs.
Huron	Canada	4.2	3.6	–	3.6	3.7	[154]
Ontario	Canada	3.9	2.6	–	3.1	–	[154]
Michigan	Canada	3.8	3.4	–	–	–	[154]
Sea							
Harbor	Norway	71–749		3–30	Nd	3–30	[155]
Harbor	Iceland	26–67		6–14	Nd	6–14	[155]
Harbor	Denmark	129–650		5–36	Nd	5–36	[155]
Baltic Sea		232–1149		18–59	Nd	18–59	[155]
North Sea		12–395			Nd	Nd	[156]
Black sea		33–1790		1.0–19	1.4–7.2	1.9–19	[157]

PFCs reach the aquatic environment either through their release into rivers or via wastewater discharge into receiving waters. In Table 6 are summarized occurrence of PFCs reported in different aquatic environments reported in Europe during recent years. Different studies on EEUU reported high concentrations in wastewater, in a recent study by Logannathan et al. [113], PFCs including perfluoroalkyl sulfonates (PFASs; PFOS, PFOSA, PFHxS) and perfluoroalkyl carboxylates (PFACs; PFOA, PFNA, PFDA, PFDoDA, PFUnDA) were investigated in two wastewater treatment plats (WWTPs). The first plant was located in Kentucky and it was representative of a rural area. The second plant was located in Georgia and it was representative of an urban area. PFOS was a major contaminant in samples from Kentucky (8.2–990 ng g^{-1} dry wt. in solid samples and 7.0–149 ng L^{-1} in aqueous samples), followed by PFOA (8.3–219 ng g^{-1} dry wt. in solid samples and 22–334 ng L^{-1} in aqueous samples). PFOA was the predominant contaminant in samples from the urban WWTP (7.0–130 ng g^{-1} dry wt. in solid samples and 1–227 ng L^{-1} in aqueous samples), followed by PFOS (<2.5–77 ng g^{-1} dry wt. in solid samples and 1.8–22 ng L^{-1} in aqueous samples). PFHxS, PFNA, PFDA, and PFOSA were detected in most of the samples, whereas PFUnDA and PFDoDA were detected in very few samples. Concentrations of some PFCs, particularly PFOA, were slightly higher in effluent than in influent, suggesting that biodegradation of some precursors contributes to the increase in PFOA concentrations in wastewater treatment processes. These mass loading values were similar to the values reported by Sinclair and Kannan [114] for New York plants and slightly higher than values reported for a Pacific Northwestern WWTP [115].

In Europe these quantities were even higher. Fifteen effluents from representative industry sectors (printing, electronics, leather, metals, paper, photographic and textiles) from Austria were analysed for PFOS. The PFOS levels ranged from 0–2.5 μg/L (2.5 $\mu g\,L^{-1}$ for leather, 0.120 $\mu g\,L^{-1}$ for metal, 0.140–1.2 $\mu g\,L^{-1}$ at four paper sites, 1.2 $\mu g\,L^{-1}$ for photographic, not found in textiles or electronics) [116]. Concentrations from 0.05 to 8.2 $\mu g\,L^{-1}$ were quantifies in the effluents of urban wastewater in Spain [117]. Predominantly, however, they are adsorbed to sewage sludge [118]. The use of sludge for land treatment or its disposal on dump sites leads to a remobilization of these recalcitrant compounds. Also, their polarity and mobility in water and soil allow them to reach the sea or groundwater unaffected.

Several studies have reported the presence of PFCs in surface waters. The occurrence of PFOA and PFOS in several surface waters in Germany was described in 2004 [119]. In summer 2006, the discovery of perfluorinated compounds in waters of the Arnsberg district in the North Rhein Westfalian Sauerland region caused a stir [120]. In this study, 12 different perfluorinated surfactants in German rivers (the Rhine River and its main tributaries, as well as the Moehne River), canals and drinking waters of the Ruhr catchments area are presented. Furthermore, the main contamination source was identified as an agricultural area on the upper reaches of the Moehne River, which is an important tributary of the Ruhr River. PFOA was the compound quantified in higher concentrations, it was found at 519 $ng\,L^{-1}$ in drinking water and at 4385 $ng\,L^{-1}$ in surface waters. In this case, the concentrations were higher than the highly polluted Tokyo Bay. In addition, the Möhne Reservoir is a source of drinking water.

In a survey study of contamination of surface and drinking waters around Lake Maggiore in northern Italy, PFCs were investigated in conjunction with other polar anthropogenic environmental pollutants [121].

PFOS and PFOA were identified as major PFCs being PFOS the most abundant one. PFOS was detected in two river water samples (Creek Vevera and River Strona) at concentrations >20 $ng\,L^{-1}$, and in the Lake Maggiore at concentrations around 8 $ng\,L^{-1}$. In addition, detection of some compounds such as PFOS and PFOA at high concentrations in rain water suggested that atmospheric deposition contributes to the contamination of the lake by these substances.

In this sense, different studies are examining precipitation (rainwater) to test for the atmospheric transformation of FTOHs as a source of PFOA and other perfluorocarboxylic acids (PFCAs) [122, 123].

A number of studies have been carried out in recent years in order to measure the occurrence of PFCs in marine environments. Sea water is a particularly challenging matrix because of the lower levels ($pg\,L^{-1}$, part-per-quadrillion) of PFCs in sea water. Yamashita used LC/ESI-MS/MS to carry out a global survey of PFOS, PFOA, PFHS, perfluorobutane sulfonate(PFBS), perfluorononanoic acid (PFNA), and perfluoro octane sulphonamide in sea water samples [124]. This paper also provides a nice summary of PFOS and PFOA measurements in the livers of various marine animals.

7
Industrial Chemicals (Corrosion Inhibitors and Plasticizers)

2-substituted benzothiazoles are a class of high-production-volume chemicals used as anticorrosion additives and biocides as well as vulcanization accelerators and antifungal agents in the paper and tanning industry. Owing to the wide application, they are regularly detected in the municipal wastewaters, being benzothiazole-2-sulfonate, benzothiazole and 2-hydroxybenzothiazole the most abundant, as shown by Kloepfer et al. [125, 126] (Fig. 5). The total concentration of six benzothiazoles in the wastewater of Berlin summed up to 3.4 $\mu g\,L^{-1}$ with the range of the temporal variability of 2–40% within 3 months.

Benzotriazoles are a class of corrosion inhibitors mainly used in deicing fluids and dishwashing agents. The main representatives 1H-benzotriazole and tolyltriazole are frequently found in wastewater of Swiss WWTP (10 and 1.6 $\mu g\,L^{-1}$ on average) [127] and in untreated municipal wastewater in the Berlin region with mean dissolved concentrations of 12 $\mu g\,L^{-1}$ for 1H-benzotriazole and 2.1 $\mu g\,L^{-1}$ and 1.3 $\mu g\,L^{-1}$ for 4- and 5-tolyltriazole, respectively [128].

Phthalate acid esters (PAEs) are a class of chemical compounds widely used in different industrial applications, mainly as plasticizers for polyvinyl chloride (PVC) resins, adhesives and cellulose film coatings and with minor applications in cosmetics, medical products, and insecticide carriers. They comprise a large group of compounds, several of them considered as

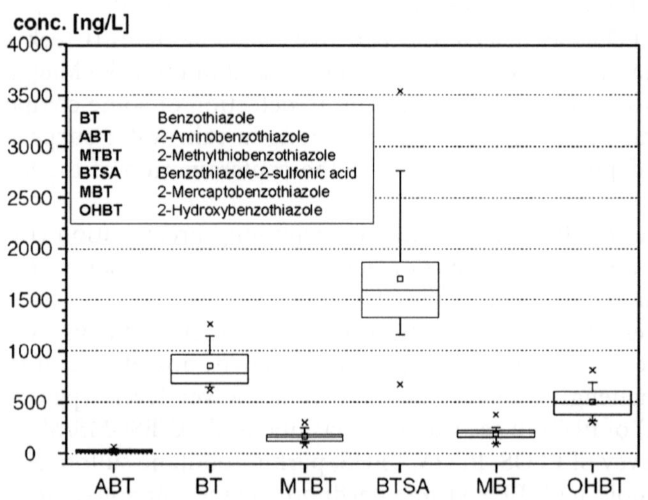

Fig. 5 Concentrations (ng/L) of the benzothiazoles in the municipal wastewater (influent to Berlin-Puhleben WWTP), summary of 20 composite samples (24 h) collected over 3 months. Adapted from [125]

priority pollutants: dimethyl (DMP), diethyl (DEP), dibutyl (DBP), butylbenzyl (BBP), di(2-ethylhexyl) (DEHP) and di-n-octyl phthalate(DnOP). The worldwide production of PAEs approximates 2.7 million metric tons a year [129] and considerable direct (production of plastic materials) and indirect emission via leaching and volatilization from plastic products after their usage, disposal and incineration, explains their ubiquity in the environment.

In all reported studies, DEHP was found to be a predominant PAE due to its high production (nearly 90% of European plasticizer use) and its physico-chemical properties (low solubility and relatively high K_{OW}). Marttineen et al. [130] reported DEHP concentrations of 98–122 $\mu g L^{-1}$ in WWTP inlet samples in Finland. Somewhat lower levels were reported by Fauser et al. [131] for inlets to WWTP in Denmark. In five Norwegian WWTP, phthalates (DEHP, BBP, DEP, DMP, and DnOP) were found in raw influent water in concentrations up to 23 $\mu g L^{-1}$ with an average of 8.0 ± 6.4 $\mu g L^{-1}$ [96]. However, contrary to other studies, DEHP was the dominant compound in only four out of 10 influent samples, while DEP was the dominating congener in the other six influent samples. The most systematic study on the occurrence of PAEs in the aquatic environment was conducted by Fromme et al. [132]. The levels of DEHP and dibutyl phthalate (DBP) were reported for 116 surface-water samples, 35 sediments from rivers, lakes and channels, 39 sewage effluents and 38 sewage sludges collected in Germany. The phthalate burden was mainly from DEHP, whilst DBP was found in minor concentrations and BBP at concentrations near the detection limit. The concentrations found ranged from 0.3–98 $\mu g L^{-1}$ (surface water), 1.7–182 $\mu g L^{-1}$ (sewage effluent), 28–154 mg/kg dw (sewage sludge) and 0.2–8.4 mg/kg (sediment). The highest concentrations found were closely related to the input of industrial wastewaters from plastic production and were limited to a few kilometers downstream of the source of contamination.

Bisphenol A (BPA) is used extensively in the production of polycarbonate, epoxy resins, flame-retardants, and many other products. Its global production is more than 1 million tons per year and a significant portion is released into surface waters [133]. In the same study, a high concentration of BPA was confirmed in waste dump water and compost water samples as well as in the liquid manure samples (61–1112 $\mu g L^{-1}$). In sewage effluents, concentrations ranged from 18 to 702 $ng L^{-1}$ and in surface waters concentrations from 0.5 to 410 $ng L^{-1}$.

8 Conclusions

The issue of emerging contaminants is closely tied to analytical capabilities. Increased sensitivity in mass spectrometry, as a result of more efficient ionization techniques and better detectors, has allowed detection of virtually any new and potentially harmful contaminant at a very low level. Consequently, a number of new or previously ignored and/or unrecognized contaminants have bean brought under scrutiny and have been detected in different environmental compartments.

Numerous papers reported on the occurrence of a wide range of emerging contaminants in the aquatic environment, being wastewater and treated wastewater (WWTP effluents) the principle source and route of their entry into the environment. However, additional monitoring studies are needed not only to confirm the presence of emerging substances in the aquatic environment but also to allow the refinement of risk assessments in combination with relevant ecotoxicological test data. In relation to the emergence of new pollutants in the environment, the integration of physical/chemical techniques, effect monitoring techniques (e.g., bioassays, functional monitoring, etc.) and ecological monitoring/assessment (community surveys) techniques play a crucial role. The main drawback of the conventional approach is target-compound monitoring, which is often insufficient to assess the environmental relevance of emerging contaminants. An integrated approach combining analytical chemistry and toxicity identification evaluation (TIE) seems to be a more appropriate way to tackle the complex problems of environmental contamination.

References

1. 6th EU Framework Programme project NORMAN (Network of reference laboratories and related organizations for monitoring and bio-monitoring of emerging environmental pollutants), Contract number 018486
2. Lienert J, Bürki T, Escher BI (2007) Water Sci Technol 56:87
3. Drillia P (2005) J Hazard Mat 122:259
4. Kuspis DA, Krenzelok EP (1996) Vet Hum Toxicol 38:48
5. Halling-Sørensen B, Nors Nielsen S, Lanzky PF, Ingerslev F, Holten Lutzhoft HC, Jørgensen SE (1998) Chemosphere 36:357
6. Daughton CG, Ternes TA (1999) Environ Health Perspect 107:907
7. Mimeault C, Woodhouse AJ, Miao XS, Metcalfe CD, Moon TW, Trudeau VL (2005) Aquat Toxicol 73:44
8. Cleuvers M (2005) Chemosphere 59:199
9. Brooks BW, Chambliss CK, Stanley JK, Ramirez A, Banks KE, Johnson RD, Lewis RJ (2005) Environ Toxicol Chem 24:464
10. Kolpin DW, Furlong ET, Meyer MT, Thurman EM, Zaugg SD, Barber LB, Buxton HT (2002) Environ Sci Technol 36:1202
11. Tolls J (2001) Environ Sci Technol 35:3397
12. Boxall ABA, Blackwell P, Cavallo R, Kay P, Tolls J (2002) Toxicol Lett 131:19
13. Jones OAH, Voulvoulis N, Lester JN (2006) Arch Environ Contam Toxicol 50:297
14. Bendz D, Paxeús NA, Ginn TR, Loge FJ (2005) J Hazard Mat 122:195
15. Alder AC, Bruchet A, Carballa M, Clara M, Joss A, Loffler D, McArdell CS, Miksch K, Omil F, Tuhkanen T, Ternes TA (2006) In: Ternes TA, Joss A (eds) Human Pharma-

ceuticals, Hormones and Fragrances: The Challenge of Micropollutants in Urban Wastewater Management. IWA Publishing, London, p 15
16. Calamari D, Zuccato E, Castiglioni S, Bagnati R, Fanelli R (2003) Environ Sci Technol 37:1241
17. Sedlak DL, Pinkston KE (2001) Water Res Update 120:56
18. Montforts MHMM (2001) In: Kümmerer K (ed) Pharmaceuticals in the environment. Springer, Berlin Heidelberg New York, p 159
19. Castiglioni S, Bagnati R, Fanelli R, Pomati F, Calamari D, Zuccato E (2006) Environ Sci Technol 40:357
20. Miao XS, Bishay F, Chen M, Metcalfe CD (2004) Environ Sci Technol 38:3533
21. Gros M, Petroviæ M, Barceló D (2007) Environ Toxicol Chem 26:1553
22. Terzic S, Senta I, Ahel M, Gros M, Petrovic M, Barcelo D, Müller J, Knepper T, Martí I, Ventura F, Jovancic P, Jabucar D, in press, Sci Total Environ
23. Santos JL, Aparicio I, Alonso E (2007) Environ Int 33:596
24. Radjenovic J, Petroviæ M, Barceló D (2007) Anal Bioanal Chem 387:1365
25. Nakada N, Tanishima T, Shinohara H, Kiri K, Takada H (2006) Water Res 40:3297
26. Heberer T (2002) Toxicol Lett 131:5
27. Ternes TA, Stumpf M, Schuppert B, Haberer K (1998) Vom Wasser 90:295
28. Gros M, Petroviæ M, Barceló D (2006) Talanta 70:678
29. Vieno NM, Tuhkanen T, Kronberg L (2006) J Chromatogr A 1134:101
30. Nikolai LN, McClure EL, MacLeod SL, Wong CS (2006) J Chromatogr A 1131:103
31. Escher BI, Bramaz N, Richter M, Lienert J (2006) Environ Sci Technol 40:7402
32. Watkinson AJ, Murby EJ, Costanzo SD (2007) Water Res 41:4164
33. Clara M, Strenn B, Gans O, Martinez E, Kreuzinger N, Kroiss H (2005) Water Res 39:4797
34. Jones OAH, Voulvoulis N, Lester JN (2001) Environ Technol 22:1383
35. Jjemba PK (2002) Agriculture, Ecosyst Environ 93:267
36. Clara M, Strenn N, Kreuzinger N (2004) Water Res 38:947
37. Gómez MJ, Martínez Bueno MJ, Lacorte S, Fernández-Alba AR, Agúera A (2007) Chemosphere 66:993
38. Lishman L, Smyth SA, Sarafin K, Kleywegt S, Toito J, Peart T, Lee B, Servos Beland M, Seto P (2006) Sci Total Environ 367:544
39. Vieno NM, (2005) Environ Sci Technol 39:8220
40. Tauxe-Wuersch A, De Alencastro LF, Grandjean D, Tarradellas J (2005) Water Res 39:1761
41. Oleksy-Frenzel J, Wischnack S, Jekel M (2000) Fresenius J Anal Chem 366:89
42. Andreozzi R, Raffaele M, Nicklas P (2003) Chemosphere 50:1319
43. Richardson ML, Bowron JM (1985) J Pharm Pharmacol 37:1
44. Khan SJ, Ongerth JE (2004) Chemosphere 54:355
45. Henschel KP, Wenzel A, Diedrich M, Fliedner A (1997) Regulat Toxicol Pharmacol 25:220
46. Ternes TA (1998) Water Res 32:3245
47. EMEA, In: Doc. Ref. EMEA/CHMP/SWP/4447/00; 2006; accessible at http:// www.emea.eu.int/pdfs/human/swp/444700en.pdf ieei, last visit to website: 10.08.2006
48. Kuster M, Lopez de Alda MJ, Rodriguez-Mozaz S, Barceló D (2007) In: Petrovic M, Barceló D (eds) Comprehensive Analytical Chemistry (Analysis, Fate and Removal of Pharmaceuticals in the Water Cycle), vol 50. Elsevier, Amsterdam, p 219
49. Schubert W, Cullberg G, Edgar B, Hedner T (1994) Maturitas 20:155
50. Kuhl H (1990) Maturitas 12:171
51. Turan A (1996) Umweltbundesamt, Berlin, p TEXTE 3/96

52. Ying G-G, Kookana RS, Ru Y-J (2002) Environ Int 28:545
53. Martindale (1982) The extra Pharmacopoeia. The Pharmaceutical Press, London
54. Petrovic M, Eljarrat E, de Alda MJL, Barcelo D (2004) Anal Bioanal Chem 378:549
55. Hanselman TA, Graetz DA, Wilkie AC, Szabo NJ, Diaz CS (2006) J Environ Qual 35:695
56. Lopez de Alda MJ, Barcelo D (2001) Fresenius J Anal Chem 371:437
57. Johnson AC, Williams RJ (2004) Environ Sci Technol 38:3649
58. Baronti C, Curini R, D'Ascenzo G, Di Corcia A, Gentili A, Samperi R (2000) Environ Sci Technol 34:5059
59. Ternes TA, Kreckel P, Mueller J (1999) Sci Total Environ 225:91
60. Desbrow C, Routledge EJ, Brighty GC, Sumpter JP, Waldock M (1998) Environ Sci Technol 32:1549
61. Belfroid AC, Van der Horst A, Vethaak AD, Schafer AJ, Rijs GBJ, Wegener J, Cofino WP (1999) Sci Total Environ 225:101
62. Beck M, Radke M (2006) Chemosphere 64:1134
63. Cui C, Ji S, Ren H (2006) Environ Monitor Assess 121:407
64. D'Ascenzo G, Di Corcia A, Gentili A, Mancini R, Mastropasqua R, Nazzari M, Samperi R (2003) Sci Total Environ 302:199
65. Gomes RL, Birkett JW, Scrimshaw MD, Lester JN (2005) Int J Environ Anal Chem 85:1
66. Koh YKK, Chiu TY, Boobis A, Cartmell E, Lester JN, Scrimshaw MD (2007) J Chromatogr A 1173:81
67. Komori K, Tanaka H, Okayasu Y, Yasojima M, Sato C (2004) Water Sci Technol 50:93
68. Laganà A, Bacaloni A, De Leva I, Faberi A, Fago G, Marino A (2004) Anal Chim Acta 501:79
69. Lee H-B, Peart TE, Svoboda ML (2005) J Chromatogr A 1094:122
70. Li Z, Wang S, Alice Lee N, Allan RD, Kennedy IR (2004) Anal Chim Acta 503:171
71. Lishman L, Smyth SA, Sarafin K, Kleywegt S, Toito J, Peart T, Lee B, Servos M, Beland M, Seto P (2006) Sci Total Environ 367:544
72. Nasu M, Goto M, Kato H, Oshima Y, Tanaka H (2001) Water Sci Technol 43:101
73. Quintana JB, Carpinteiro J, Rodriguez I, Lorenzo RA, Carro AM, Cela R (2004) J Chromatogr A 1024:177
74. Reddy S, Iden CR, Brownawell BJ (2005) Anal Chem 77:7032
75. Roda A, Mirasoli M, Michelini E, Magliulo M, Simoni P, Guardigli M, Curini R, Sergi M, Marino A (2006) Anal Bioanal Chem 385:742
76. Servos MR, Bennie DT, Burnison BK, Jurkovic A, McInnis R, Neheli T, Schnell A, Seto P, Smyth SA, Ternes TA (2005) Sci Total Environ 336:155
77. Zuehlke S, Duennbier U, Heberer T (2005) J Sep Sci 28:52
78. Johnson AC, Belfroid A, Di Corcia A (2000) Sci Total Environ 256:163
79. Nakada N, Tanishima T, Shinohara H, Kiri K, Takada H (2006) Water Res 40:3297
80. Clara M, Kreuzinger N, Strenn B, Gans O, Kroiss H (2005) Water Res 39:97
81. Castiglioni S, Bagnati R, Fanelli R, Pomati F, Calamari D, Zuccato E (2006) Environ Sci Technol 40:357
82. Ternes TA, Stumpf M, Mueller J, Haberer K, Wilken R-D, Servos M (1999) Sci Total Environ 225:81
83. Huang CH, Sedlak DL (2001) Environ Toxicol Chem 20:133
84. EMCDDA (2007) European Monitoring Centre for Drugs and Drug Addiction, Lisbon, http://www.emcdda.europa.eu/html.cfm/index875EN.html#42164
85. UNODC (2007) United Nations Office on Drugs and Crime, Vienna, http://www.unodc.org/pdf/research/wdr07/WDR_2007_executive_summary.pdf

86. Huerta-Fontela M, Galcerán MT, Ventura F (2007) Anal Chem 79:3821
87. Castiglioni S, Zuccato E, Crisci E, Chiabrando C, Fanelli R, Bagnati R (2006) Anal Chem 78:8421
88. Bones J, Thomas KV, Paull B (2007) J Environ Monit 9:701
89. Zucatto E, Chiabrando C, Castiglioni S, Calamari D, Bagnati R, Schiarea S, Fanelli R (2005) Environ Health: Global Access Sci Source 4:1
90. Hummel D, Löffler D, Fink G, Ternes TA (2006) Environ Sci Technol 40:7321
91. Jones-Lepp TL, Alvarez DA, Petty JD, Huckins JN (2004) Arch Environ Contam Toxicol 47:427
92. Postigo C, López de Alda MJ, Barceló D (2007) Anal Chem, in press
93. Pizzolato TM, Lopez de Alda MJ, Barceló D (2007) Trends Anal Chem 26:609
94. Daughton CG (2001) In: Pharmaceuticals and personal care products in the environment: Scientific and regulatory issues. Daughton CG, Jones-Lepp TL (eds) ACS Symposium Series 791. The American Chemical Society, Washington DC, p 116
95. Knepper TP, Eichhorn P, Bonnington LS (2003) Aerobic degradation of surfactants. In: Knepper TP, Barceló D, de Voogt P (eds) Analysis and Fate of Surfactants in the Aquatic Environment. Elsevier, Amsterdam, The Netherlands, p 525
96. Vogelsang C, Grung M, Jantsch TG, Tollefsen KE, Liltved H (2006) Water Res 40:3559
97. Giger W, Alder AC, Ahel M, Schaffner C, Reiser R, Albrecht A, Lotter AF, Sturm M (2002) Proc 1st SedNet Workshop on Chemical Analysis and Risk Assessment of Emerging Contaminants in Sediments and Dredged Material, Barcelona, Spain, p 183
98. Gonzalez S, Petrovic M, barcelo D (2004) J Chromatogr A 1052:111
99. Kvestak R, Terzic S, Ahel M (1994) Mar Chem 46:89
100. Jobling S, Sheahan D, Osborne JA, Matthiessen P, Sumpter JP (1996) Environ Toxicol Chem 15:194
101. Croatian Ordinance on Maximum Permissible Concentrations of Hazardous Contaminants in Waters and Coastal Sea. Narodne novine, No 2, 1984
102. Ahel M, Giger W, Koch M (1994) Water Res 28:1131
103. Vieira VM (2005) Perfluorinated compounds (PFCs). In: Health effects review. International Joint Commission: Department of Environmental Health, Boston University School of Public Health, Boston, MA (http://www.ijc.org/ rel/pdf/health_effects_spring 2005.pdf, last accessed 28th April 2006)
104. Ueno D, Darling C, Alaee M, Campbell L, Pacepavicius G, Teixeira C, Muir D (2007) Environ Toxicol Chem 41:841
105. Armitage J, Cousins IT, Buck RC, Prevedouros K, Russell MH, MacLeod M, Korzeniowski SH (2006) Environ Sci Technol 40:6969
106. Tomy GT, Budakowski W, Halldorson T, Helm PA, Stern GA, Friesen K, Pepper K, Tittlemier SA, Fisk A (2004) Environ Sci Technol 38:6475
107. Lau C, Thibodeaux JR, Hanson RG, Narotsky MG, Rogers JM, Lindstrom AB, Strynar MJ (2006) Toxicol Sci 90:510
108. Kennedy GL, Butenhoff JL, Olsen GW, O'Connor JC, Seacat AM, Perkins RG, Biegel LB, Murphy SR, Farrar DG (2004) Crit Rev Toxicol 34:351
109. Richardson SD (2007) Anal Chem 79:4295
110. Council of the European Communities (1996) Off J Eur Commun L 257:0026–0040
111. Prevedouros K, Cousins IT, Buck RC, Korzeniowski SH (2006) Environ Toxicol Chem 40:32
112. Wania F (2007) Environ Toxicol Chem 41:4529
113. Loganathan BG, Sajwan KS, Sinclair E, Kumar KS, Kannan K (2007) Water Res 41:4611

114. Sinclair E, Kannan K (2006) Environ Sci Technol 40:1408
115. Schultz MM, Higgins CP, Huset CA, Luthy RG, Barofsky DF, Field JA (2006) Environ Sci Technol 40:7350
116. Hohenblum P, Scharf S, Sitka A (2003) Vom Wasser 101:155
117. Alzaga R, Bayona JM (2004) J Chromatogr A 1042:155
118. Schröder HF (2003) J Chromatogr A 1020:131
119. Lange FT, Schmidt CK, Metzinger M, Wenz M, Brauch HJ (2004) Poster at the SETAC-Meeting in Prague (CZ), April 18th–22th
120. Skutlarek D, Exner M, Färber H (2006) Environ Sci Pollut Res 13:299
121. Loos R, Wollgast J, Huber T, Hanke G (2007) Anal Bioanal Chem 387:1469
122. Scott BF, Moody CA, Spencer C, Small JM, Muir DCG, Mabury SA (2006) Environ Sci Technol 40:6405
123. Scott BF, Spencer C, Mabury SA, Muir DCG (2006) Environ Sci Technol 40:7167
124. Yamashita N, Kannan K, Taniyasu S, Horii Y, Petrick G, Gamo T (2005) Mar Pollut Bull 51:658
125. Kloepfer A, Gnirss R, Jekel M, Reemtsma T (2004) Water Sci Technol 50:203
126. Kloepfer A, Jekel M, Reemtsma T (2005) Enviorn Sci Technol 39:3792
127. Voutsa D, Hartmann P, Schaffner C, Giger W (2006) Environ Sci Pollut Res 13:333
128. Weiss S, Jakobs J, Reemtsma T (2006) Environ Sci Technol 40:7193
129. Baurer MJ, Herman R (1997) Sci Total Environ 208:49
130. Marttinen SK, Kettunen RH, Sormunen KM, Rintala JA (2003) Water Res 37:1385
131. Fauser P, Vikelsoe J, Sorensen PB, Carlsen L (2003) Water Res 37:1288
132. Fromme H, Küchler T, Otto T, Pilz K, Müller J, Wenzel A (2002) Water Res 36:1429
133. Staples CA, Dorn PB, Klecka GM, O'Block ST, Harris LR (1998) Chemosphere 36:2149
134. Carballa M, Omil F, Lema JM, Llompart M, García-Jares C, Rodríguez I, Gómez M, Ternes T (2004) Water Res 38:2918
135. MacLeod SL, Sudhir P, Wong CS (2007) J Chromatogr A 1170:23
136. Ternes TA, Hirsch R (2000) Environ Sci Technol 34:2741
137. Li HQ, Jiku F, Schröder HF (2000) J Chromatogr A 889:155
138. Clara M, Scharf S, Scheffknecht, Gans O (2007) Water Res 41:4339
139. Di Corcia A, Cavallo R, Crescenzi C, Nazzari M (2000) Environ Sci Technol 34:3914
140. Crescenzi C, Di Corcia A, Samperi R (1995) Anal Chem 67:1797
141. Eichhorn P, Petroviæ M, Barceló D, Knepper TP (2000) Vom Wasser 95:245
142. Planas C, Guadayol JM, Doguet M, Escalas A, Rivera J, Caixach J (2002) Water Res 36:982
143. Castillo M, Martinez E, Ginebreda A, Tirapu L, Barceló D (2000) Analyst 125:1733
144. Ahel M, Molnar E, Ibric S, Giger W (2000) Water Sci Technol 42:15
145. Ahel M, Giger W (1998) Am Chem Soc Nat Meeting Extended Abst 38:276
146. De Voogt P, Kwast O, Hendriks R, Jonkers CCA (2000) Analysis 28:776
147. Castillo M, Alonso MC, Riu J, Barceló D (1999) Environ Sci Technol 33:1300
148. Tanghe T, Devriese G, Verstraete W (1999) J Environ Qual 28:702
149. Di Corcia A, Samperi R (1994) Environ Sci Technol 28:850
150. González-Barreiro C, Martínez-Carballo E, Sitka A, Scharf S, Gans O (2006) Anal Bioanal Chem 386:2123
151. McLachlan M, Holstrom K, Andursberger M (2007) Environ Sci Technol 41:7260
152. Senthikumar K, Ohi E, Sajwan K, Takasuga T, Kannan K (2007) Bull Environ Contam Toxicol 79:427
153. Rostkowski O, Yamashita N, Man Ka So I, Taniyasu S, Kwan Sing Lam P, Falandysz J, Tae Lee K, Kyu Kim S, Seong Khim J, Hyeon Im S, Newsted JL, Jones P, Kannan K, Giesy JP (2006) Environ Toxicol Chem 25:2374

154. Furdui VI, Stock NL, Ellis DA, Butt CM, Whittle DM, Crozier PW, Reiner EJ, Mabury SA (2007) Environ Sci Technol 41:1554
155. Tanabe S, Madhusree B, Ozturk AA, Tatsukawa R, Miyazaki N, Ozdamar E, Aral O, Samsun O, Ozturk B (1997) Mar Pollut Bull 34:338
156. Houde M, Bujas TAD, Small J, Wells RS, Fair PA, Bossart GD, Solomon KR, Muir DCG (2006) Environ Sci Technol 40:4138
157. Inneke K, de Vijver V, Holsbeek L, Das K, Blust R, Joiris C, De Coen W (2007) Environ Sci Technol 41:315

Hdb Env Chem Vol. 5, Part S/1 (2008): 37–104
DOI 10.1007/698_5_102

Published online: 8 March 2008

Analysis of Emerging Contaminants of Municipal and Industrial Origin

Meritxell Gros[1] (✉) · Mira Petrovic[1,2] · Damià Barceló[1]

[1]Department of Environmental Chemistry, IIQAB-CSIC, c/Jordi Girona 18–26, 08034 Barcelona, Spain
megqam@cid.csic.es

[2]Institució Catalana de Recerca i Estudis Avançats (ICREA), Passeig Lluís Companys 23, 80010 Barcelona, Spain

Abstract Besides recognized pollutants, numerous other chemicals are continuously released into the environment as a result of their use in industry, agriculture, consumer goods or household activities. The presence of these substances, known as emerging contaminants, has become an issue of great concern within the scientific community during the last few years. For this reason, the availability of sensitive, accurate and reliable analytical techniques is essential in order to assess their occurrence, removal and fate in the environment.

In this chapter, the state of the art of the analytical techniques used to determine a wide range of emerging contaminants in several environmental matrices will be overviewed.

Keywords Emerging contaminants · Instrumental analysis · Sample preparation techniques

Abbreviations

ADBI	4-Acetyl-1,1-dimethyl-6-*tert*-butylindane
AED	Atomic emission detector
AHMI	6-Acetyl-1,1,2,3,3,5-hexamethylindane
AHTN	7-Acetyl-1,1,3,4,4,6-hexamethyl-1,2,3,4-tetrahydronaphthalene
AP	Alkylphenol
APCI	Atmospheric pressure chemical ionization
APEC	Alkylphenoxy carboxylate
APEO	Alkylphenol ethoxylate
APPI	Atmospheric pressure photoionization
ATII	5-Acetyl-1,1,2,6-tetramethyl-3-isopropylindane
BSA	*N*,*O*-Bis(trimethylsilyl)-acetamide
BSTFA	*N*,*O*-Bis(trimethylsilyl)-trifluoroacetamide
BTEX	Benzene, toluene, ethylbenzene and xylenes
CAPEC	Dicarboxylated alkylphenoxy ethoxylate
CAR	Carboxen
CDEA	Coconut diethanolamide
CID	Collision-induced dissociation
CLLE	Continuous liquid–liquid extraction
CSIA	Compound-specific stable isotope analysis
CW	Carbowax
DAI	Direct aqueous injection
DEET	*N*,*N*-Diethyl-*m*-toluamide
DI-SPME	Direct solid-phase microextraction
DMIP	Dummy molecularly imprinted polymer
DPMI	6,7-Dihydro-1,1,2,3,3-pentamethyl-4-(5*H*)-indanone
DVB	Divinylbenzene
ECD	Electron capture detector

EI	Electron impact
ELISA	Enzyme-linked immunosorbent assay
EPA	Environmental Protection Agency
ESI	Electrospray ionization
EU	European Union
FAS	Fluorinated alkyl substance
FID	Flame ionization detector
F NMR	Fluorine nuclear magnetic resonance
FTOH	Fluorotelomer alcohol
GC	Gas chromatography
GCB	Graphitized carbon black
GC×GC	Comprehensive two-dimensional gas chromatography
GC-MS	Gas chromatography–mass spectrometry
GPC	Gel permeation chromatography
HHCB	1,2,4,6,7,8-Hexahydro-4,6,6,7,8,8-hexamethylcyclopenta-γ-2-benzopyrane
HLB	Hydrophilic–lipophilic balanced
HPLC	High-performance liquid chromatography
HS	Headspace
HSGC	Headspace gas chromatography
HS-SPME	Headspace solid-phase microextraction
IA	Immunoaffinity
IDA	Information-dependent acquisition
IPPC	Integrated Prevention and Control of the Contamination Directive
KOH	Potassium hydroxide
LAS	Linear alkyl sulphonate
LC	Liquid chromatography
LC/ESI-MS	Liquid chromatography–electrospray mass spectrometry
LLE	Liquid–liquid extraction
MAE	Microwave-assisted extraction
MCF	Methyl chloroformate
MCX	Mixed-mode cation exchange
MIMS	Membrane-introduction mass spectrometry
MIP	Molecularly imprinted polymer
MMLLE	Microporous membrane liquid–liquid extraction
MRM	Multiple reaction monitoring
MSPD	Matrix solid-phase dispersion
MSTFA	*N*-Methyl-*N*-trimethylsilyltrifluoroacetamide
MTBE	Methyl *tert*-butyl ether
MTBSTFA	*N*-(*tert*-Butyldimethylsilyl)-*N*-methyltrifluoroacetamide
NCI	Negative chemical ionization
NI	Negative ionization
NP	Normal phase
NPEC	Nonylphenoxy carboxylate
OECD	Organization for Economic Co-operation and Development
PA	Polyacrylate
PAH	Polycyclic aromatic hydrocarbon
PAM-MS	Purge-and-membrane inlet mass spectrometry
PBDE	Polybrominated diphenyl ether
PCB	Polychlorinated biphenyl
PCI	Positive chemical ionization

PCP	Personal care product
PDMS	Polydimethylsiloxane
PEEK	Polyetheretherketone
PFA	Pentafluoropropionic acid anhydride
PFDA	Perfluorodecanoic acid
PFO	Perfluorooctane sulphonate
PFOA	Perfluorooctanoate
PI	Positive ionization
PID	Photoionization detector
PLE	Pressurized-liquid extraction
PPY	Polypyrrole
PTFE	Polytetrafluoroethylene
PTV	Programmable temperature vaporization
P&T	Purge and trap
Q-LIT	Quadrupole–linear ion trap
QqQ	Triple quadrupole
Q-TOF	Quadrupole–time of flight
RAM	Restricted access material
RIA	Radioimmunoassay
RP	Reversed phase
SAX	Strong anion exchange
SEC	Size-exclusion chromatography
SFE	Supercritical-fluid extraction
SIM	Selected ion monitoring
SNUR	Significant new use rule
SPE	Solid-phase extraction
SPME	Solid-phase microextraction
SRM	Selected reaction monitoring
TBA	*tert*-Butyl alcohol
TBBPA	Tetrabromobisphenol A
TBF	*tert*-Butyl formate
TBS	*tert*-Butyldimethylsilyl
TFC	Turbulent flow chromatography
TMS	Trimethylsilyl
TMS-DEA	*N*,*N*-Diethyltrimethylamine
TrBA	Tri-*n*-butylamine
UPLC	Ultra-performance liquid chromatography
UV	Ultraviolet
VOC	Volatile organic compound
WAX	Mixed mode weak anion exchange
WWTP	Wastewater treatment plant

1 Introduction

During the last three decades, the impact of chemical pollution has focused almost exclusively on the conventional "priority" pollutants, which have long been recognized as posing risks to human health, due to their toxicity, carcinogenic and mutagenic effects, and their persistence in the environment. Legislation and long-established standards and certified analytical methods, set by the Environmental Protection Agency (EPA) and the International Organization for Standardization (ISO), are already available for the determination of these priority pollutants. Besides recognized contaminants, numerous other chemicals are continuously released into the environment as a result of their use in industry, agriculture, consumer goods or household activities. The identification, analysis and characterization of the risks posed by these substances, classified as the so-called emerging contaminants, has focused attention and awakened concern among the scientific community during the last few years. This group of compounds, including pharmaceuticals and personal care products, surfactants, gasoline additives, fire retardants and fluorinated organic compounds, among others, is still unregulated. These contaminants may be candidates for future regulation, depending on research on their potential health effects and monitoring data regarding their occurrence.

Several studies have demonstrated that wastewater treatment plants (WWTPs) are major contributors to the presence of emerging contaminants in the environment. As these substances are used in everyday life, they are continuously introduced into the aquatic media via sewage waters mainly through industrial discharges (surfactants, fire retardants), excretion (pharmaceuticals, hormones and contraceptives, personal care products) or disposal of unused or expired substances [1]. Methyl *tert*-butyl ether (MTBE) and other gasoline additives also enter the aquatic environment due to anthropogenic activities, mainly via accidental spills and leakage of corroded tanks at gasoline stations or refineries.

Due to their continuous introduction into the environment, emerging contaminants can be considered as "pseudo-persistent" pollutants, which may be able to cause the same exposure potential as regulated persistent pollutants, since their high transformation and removal rates can be compensated by their continuous input into the environment [2]. Consequently, there is a growing need to develop reliable analytical methods, which enable their rapid, sensitive and selective determination in different environmental compartments at trace levels.

This chapter aims to overview the state of the art of the most recent analytical methodologies developed in the last few years for the analysis of emerging contaminants in environmental samples, using advanced chromatographic techniques and detection systems. Since it is impossible to cover all analytes, we have just focused our attention on selected classes of contaminants, which are currently the most widely studied and ubiquitous in the environment. Trends in sample preparation and instrumental analysis for each group of compounds will be described.

2 Sampling and Sample Preparation

Sample preparation is one of the most important steps within an analytical methodology. Selectivity of stationary phases used for the isolation and pre-concentration of target compounds is a key parameter to take into account when analysing emerging contaminants at trace levels from complex environmental samples, since the reduction of co-extracted compounds results in a better sensitivity, achieving lower limits of detection. In the following section, a summary of the trends in stationary phases and materials used for the analysis of emerging contaminants in both aqueous and solid samples will be described.

2.1 Sampling Strategies

Generally, to determine surface waters (river, lake, sea) grab samples are used, whereas for wastewaters composite samples are often collected over sampling periods of 6 h to several days. Some studies reported that the addition of 1% of formaldehyde to water samples prevents degradation of target compounds until analysis. Before sample enrichment, water samples are filtered through glass fibre or cellulose filters. Depending on the nature of the water sample (wastewater, surface water or seawater) and its organic matter content, different pore size filters are used.

In the case of sediments or soil samples, depending on the objective of the study (determination of vertical distribution profiles or concentrations in a surface layer), either core or grab samples are taken. Usually, water is removed and then the solid matrix is stored in the dry state. Removal of water from the sediments before extraction was found to be crucial in obtaining good recoveries [3]. Freeze-drying is an accepted and commonly used procedure for drying solid matrices, but it is not known how this affects the levels of target compounds measured, especially those that are relatively volatile [4].

When small fish, mussels or other bivalves are analysed, several individual species are homogenized to form a pool of tissues, from which sub-samples are taken for extraction. Removal of water is also generally performed by freeze-drying [5].

However, for aqueous matrices, grab samples may not be representative and moreover, a relatively large number of samples must be taken from a given location over the entire duration of sampling [6]. Therefore, a good alternative to overcome this problem could be the use of passive samplers. These devices are based on the free flow of analyte molecules from the sampled medium to a collecting one, as a result of a difference in chemical potentials of the analyte between the two media. Although they have only been applied for the determination of some organic pollutants and pesticides, their application in aqueous and gaseous phases is constantly increasing [6–10].

In passive samplers, the concentration of the analyte is integrated over the whole exposure time, making it immune to accidental or extreme variations of pollutant concentrations [6]. Other advantages against grab sampling are that decomposition of the sample during transport and storage is minimized and that passive sampling and/or extraction methods are simple to perform as, after the isolation and/or enrichment step, no further sample preparation is usually required [6]. Devices used today are based on diffusion through a well-defined diffusion barrier or permeation through a membrane, the former being the most popular ones.

2.2 Analysis of Emerging Contaminants in Water Samples

Extraction of target compounds from water matrices is generally achieved by solid-phase extraction (SPE) and solid-phase microextraction (SPME). For SPE, several stationary phases can be used, ranging from mixtures of different polymers (such as divinylbenzene–vinylpyrrolidone) to octadecylsilica (C_{18}) or more selective tailor-made materials, such as immunosorbents, molecularly imprinted polymers (MIPs) and restricted access materials (RAMs).

The use of tailor-made materials is very useful when performing single group analysis, as they enhance the selectivity for the compounds of interest in the sample preparation process, reducing the amount of co-extracted material and, as a result, increasing the sensitivity. However, when the aim of the analytical methodology is to analyse a wide spectrum of compounds with different physico-chemical properties, polymeric or C_{18} sorbents are the most recommended ones.

The use of automated on-line systems, which integrate extraction, purification and detection, has increased over the past several years. One option is on-line coupling of SPE and LC, utilizing special sample preparation units, such as PROSPEKT (Spark Holland) and OSP-2 (Merck). This technique has been successfully applied to the analysis of pesticides, estrogens and progestogens in water samples [11–17]. Similarly, on-line coupling of SPE and SPME to GC is a promising approach with good prospects [18, 19].

2.2.1 Immunosorbents

The immunosorbents, such as polyclonal antibodies, are immobilized on silica-based supports, activated Sephadex gels, synthetic polymers, sol/gel materials, cyclodextrins, or RAMs and packed into cartridges or pre-columns [20, 21]. Immunoaffinity extraction coupled with LC/ESI-MS has been used for the analysis of pesticides [12, 22–24] and β-estradiol and estrone in wastewater [25]. Immunosorbents have also the potential to be applied to the determination of drugs in aqueous samples. In fact, most on-line immunosorbent applications correspond to pharmaceutical and biomedical trace analysis [26]. Therefore, a high number of pharmaceuticals [27, 28] and hormones [29, 30] have been determined in biological samples using immunoaffinity SPE coupled to on-line LC-MS. With these materials, humic and fulvic acids are not co-extracted and thus no further clean-up is necessary. Moreover, cross-reactivity of the antibody can be advantageous, because it not only extracts a determined substance, but also all compounds within a given class, being then separated and quantified individually by coupling with chromatographic techniques [31].

2.2.2 Molecularly Imprinted Polymers (MIPs)

During the last few years, MIPs have appeared as new selective sorbents for SPE of organic compounds in complex materials [32, 33]. Both on-line and off-line MIP-SPE protocols have been developed to determine organic pollutants in environmental waters, mainly pesticides and hormones [34–39].

Molecular imprinting is a rapidly developing technique for the preparation of polymers having specific molecular recognition properties [40–43]. First, the template and the monomer form a stable template–monomer complex prior to polymerization. Then the complex is polymerized in the presence of a cross-linking agent. The resulting MIPs are matrices possessing microcavities with a three-dimensional structure complementary in both shape and chemical functionality to that of the template [44, 45]. After polymerization, the template, which consists of one of the target analytes or related analogues, is removed, generating specific binding sites. Then, the polymer can be used to selectively rebind the template molecule, the analyte or structurally related analogues. The specific binding sites in MIPs are formed by covalent or, more commonly, non-covalent interactions between the imprinting template and the monomer [32].

Apart from their high selectivity for target compounds, MIPs possess other advantages, such as low cost, high stability, ability to be reused without loss of activity, high mechanical strength, durability to heat and pressure and applicability in harsh chemical media [46, 47].

MIPs can be prepared in a variety of physical forms, but the conventional approach is to synthesize the MIP in bulk, grind the resulting polymer and sieve the particles into the desired size ranges [48, 49]. However, this method is tedious and time-consuming, often produces particles that are irregular in size and shape and some interaction sites are destroyed during grinding. In order to overcome these problems, alternative methods have been developed, such as using multi-step swelling procedures, suspension and precipitation polymerization, respectively, to obtain uniform spherical particles [50–55].

In MIP-SPE processes, the sample medium, during the loading step, has an important influence on the recognition properties of the MIP. If the analyte of interest is presented in an aqueous medium, the analyte and other interfering compounds are retained non-specifically on the polymer. Therefore, to achieve the selectivity desired, a clean-up step using organic solvents is required prior to elution [32].

One of the main disadvantages of MIP-SPE is the difficulty in removing the entire template molecule, even after extensive washing, and therefore a leakage of template molecule can occur, which is an obstacle in the determination of target compounds. To overcome this problem, a structural analogue of the target molecule can be imprinted to make a "dummy molecularly imprinted polymer" (DMIP), distinguishing then any leakage of target compound [56].

2.2.3 Restricted Access Materials (RAMs)

RAMs are a class of SPE materials that possess a biocompatible surface and a pore size that restricts big molecules from entering the interior extraction phase based on size [26]. Simultaneously, an extraction phase located on the inner pore surface is responsible for isolation of the low molecular weight compounds [26]. Koeber et al. [57] applied this approach in combination with MIP and used an on-line mode to analyse pesticides from environmental samples. There are various references reporting the use of RAMs for direct injection of biological samples [58–60], but few applications have been reported for environmental matrices.

2.2.4 Solid-Phase Microextraction (SPME)

Several reviews have been devoted to the application of SPME in environmental analysis [6, 61–66]. SPME is a simple and effective adsorption/absorption and desorption technique which eliminates the need for solvents and combines sampling, isolation and enrichment in one step [66]. Depending on the analyte and matrix, SPME of water samples can be performed in different modes: direct-immersion extraction (for less volatile compounds and relatively clean samples), headspace extraction (for more volatile compounds and dirtier samples), membrane-protected SPME (for the extraction of analytes in heavily polluted samples), in-tube SPME [5, 67] and thin-film microextraction (use of a thin sheet of PDMS membrane) [68].

In-tube SPME has been applied for the determination of a variety of environmental pollutants [69–75] and is based on the use of a fused-silica capillary column as the extraction device. Target analytes in aqueous matrices are directly extracted and concentrated by the coating in the capillary column by repeated withdrawal and expulsion of the sample solution, and can be directly transferred to LC or GC columns for analysis.

The major part of SPME applications has been developed for GC, as the coupling to HPLC is more complex and requires specifically designed interfaces to desorb analytes from the fibres and also because not all fibres can be used for LC, due to solubility and swelling of the fibre coatings in organic solvents [5].

Several fibre coatings are commercially available for the analysis of non-polar organic compounds, such as BTEX, PAHs and pesticides, and polar compounds like phenols, alcohols, etc. [66], including polydimethylsiloxane (PDMS), polyacrylate (PA), divinylbenzene (DVB), Carboxen (CAR) and Carbowax (CW). On the other hand, a polypyrrole (PPY) coating is used to extract polar or ionic analytes [67], which is mainly addressed to the coupling of SPME to LC.

Another way to determine polar compounds by SPME is presented by SPME derivatization, which includes three different approaches: in-coating, direct or on-fibre derivatization. The difference between these techniques is that while in direct derivatization, the derivatizing agent is first added to the sample vial and the derivatives are then extracted by the SPME fibre coating, for on-fibre derivatization, the derivatizing agent is loaded on the fibre, which is subsequently exposed to the sample and extracted [66]. This approach is now widely used for the analysis of organic pollutants in the environment, such as acidic herbicides [76, 77], and has been recently reviewed by Stashenko [78] and Dietz [79].

2.3 Analysis of Emerging Contaminants in Solid Samples and Biota

2.3.1 Extraction Techniques

Organic contaminants present in solid environmental samples, such as sediments, soils, sludge and biota, are determined by exhaustive extraction with appropriate solvents. Liquid–liquid extraction (LLE), Soxhlet, sonication, pressurized-liquid extraction (PLE), microwave-assisted extraction (MAE) and supercritical-fluid extraction (SFE) are the techniques most commonly used [5]. Also methods based on HS-SPME have been developed to determine volatile and semi-volatile compounds.

Soxhlet has been widely used, as it is considered as the reference method, is inexpensive and is easy to handle. However, new trends are focused on the use of "low-solvent, low-time and low-cost" techniques, amenable to automation, such as PLE, MAE and SFE. These techniques use elevated temperature and pressure, which results in improved mass transfer of the analytes and, consequently, increased extraction efficiency. SFE and MAE are not suitable for highly polar organic compounds or matrices with high water content. Therefore, nowadays PLE, also termed accelerated solvent extraction, is the preferred technique, because it is automated, it consumes low amounts of solvent and because older extraction procedures can be easily adapted. However, it offers some disadvantages, such as its cost, as commercial PLE equipment may be expensive and, moreover, some thermolabile compounds may suffer degradation. A good alternative to PLE would be MAE, as it is more affordable, fast and consumes little solvent, but extracts need to be filtered and microwave heating is uneven and restricted to matrices that adsorb this radiation. SFE with solid-phase trapping has been used for different groups of organic pollutants. Although good results and unique improved selectivity were obtained for selected applications, the method did not find acceptance. This is because the extraction conditions depend on the sample, requiring complicated optimization procedures [5, 80].

2.3.2 Extract Clean-up and Purification

Due to the complexity of samples and the exhaustive extraction techniques used, a substantial number of interfering substances present in the matrix are found in the extracts. Therefore, a clean-up and purification step after extraction is indispensable to remove these compounds and enhance selectivity, in order to reduce ion-suppression effects when working with ESI-MS detection and to improve the separation of analytes from impurities.

2.3.2.1 Solid Samples

The conventional approach used is based on solid/liquid adsorption, using either long open columns or disposable cartridges packed with different sorbents, depending on the physico-chemical properties of the analytes of interest. Purification can be also performed by off-line SPE cartridges packed with polymeric materials, C_{18}, NH_2-, CN-modified silica or anionic exchange materials, by reversed-phase (RP) or normal-phase (NP) liquid chromatography, generally using alumina, silica or Florisil as the packing material, or size-exclusion chromatography (SEC) [5]. When high selectivity for one compound or related analogues is desired, MIPs and RAMs are also appropriate materials to use for the clean-up of crude extracts.

Purification based on two tandem SPE procedures is a widespread approach, which generally consists of the use of anionic exchange cartridges and other polymeric materials. Moreover, when extracts contain high amount of lipids and organic matter, such as sewage sludge and biota, non-destructive and destructive methods are generally used prior to instrumental analysis. The former include gel permeation and column adsorption chromatography, generally using polystyrene–divinylbenzene copolymeric columns. Other neutral adsorbents commonly used are silica gel, alumina and Florisil® [81]. Destructive lipid removal methods consist of sulphuric acid treatment, either directly to the extract or via impregnated silica columns, and saponification of extracts by heating with ethanolic KOH [82].

2.3.2.2
Biota

The analysis of biota, such as fish or mussels, could be an indicator of the water quality, as lipophilic organic contaminants tend to accumulate in the tissues with high lipid content. Isolation of organic compounds from biological tissues is a complicated and laborious task because of the nature of the matrix. Disruption of a cellular structure of biological samples results in an abundance of lipids and proteins. Extraction methods often yield high concentrations of lipids and, therefore, an exhaustive purification is required to achieve the selectivity and sensitivity desired. For this reason, treatment with sulphuric acid and saponification are frequently used for the removal of lipids prior to the purification using the same techniques as for solid samples (RP or NP, LC, SPE, SEC, MIP or RAM). However, in some cases, this step has to be avoided as some target compounds may be destroyed.

A simultaneous extraction and clean-up step was proposed by Eljarrat et al. [83] for the determination of PBDEs in fish. This methodology is based on the inclusion of alumina in the PLE cells, so that both purification and isolation of target analytes is achieved in a single step, speeding up sample preparation considerably.

Another approach to conduct simultaneous disruption and extraction of solid and semi-solid samples involves matrix solid-phase dispersion (MSPD), a technique that combines in one step extraction, concentration and clean-up by blending a small amount of sample with the selected sorbent. It has been successfully applied to the analysis of penicillins, sulphonamides, tetracycline antibiotics [5] and ionic [5, 84, 85] and non-ionic surfactants in fish and mussels.

3
Instrumental Analysis and Quantitation

3.1
Chromatographic Separation

Both gas chromatography (GC) and liquid chromatography (LC) are techniques par excellence in environmental analysis. Even though the former is more addressed to the analysis of non-polar and volatile compounds (PBDEs and MTBE), non-volatile compounds, such as pharmaceuticals, surfactants, personal care products, estrogens and others, can also be determined after a derivatization step.

3.1.1
Gas Chromatography

GC was one of the first chromatographic separation techniques to be developed, and today is still widely used and has not lost its eminence in the environmental field. The popularity of GC is based on a favourable combination of very high selectivity and resolution, good accuracy and precision, wide dynamic range and high sensitivity. Columns mainly used in GC consist of narrow-bore capillary columns [86–88].

In GC, the three most frequently used injection systems are splitless, on-column and programmable temperature vaporization (PTV). In splitless injection, the transfer of the analytes into the analytical column is controlled by the volume of the liner and by the injected volume. In on-column injection, extracts are directly injected into the column or in a glass insert fitted into a septum-equipped programmable injector kept at low temperature. Finally, PTV is a split/splitless injector which allows the sample to be introduced at a relatively low temperature, thus affording accurate and reproducible sampling. After injection, the PTV is rapidly heated to transfer the vaporized components into the capillary column.

Nowadays, headspace GC (HSGC) and comprehensive two-dimensional GC (GC×GC) have gained popularity in the environmental field. The main advantages presented by the former, against GC, is the ability to increase efficiency and drastically reduce analysis time [89]. On the other hand, GC×GC has a great capability to separate and identify organic compounds in complex environmental samples. This technique has been mainly employed for the determination of MTBE and other oxygenated and aromatic compounds in gasoline-contaminated ground waters [90] and for the determination of PBDEs [91]. In this technique, two GC separations based on distinctly different separation mechanisms are used, with the interface, called modulator, between them. Then, the effluent from the first column is separated into a large number of small fractions, and each of these is subsequently separated on the second column, which is much faster than the first separation. In principle, all kinds of stationary phases can be used in the first dimension of a GC×GC system, but generally, non-polar phases are the preferred ones. Concerning the second dimension, a variety of phases can be selected depending on the desired analyte–stationary phase interactions. However, most applications showed that the combination between a non-polar and (medium) polar phase is by far the most popular option. Concerning column size, samples are generally first separated on a 15–30 m × 0.25–0.32 mm ID × 0.1–1 μm film (d_f) column. After modulation, each individual fraction is injected onto a much shorter, narrower column, with dimensions typically 0.5–2 m × 0.1 mm ID × 0.1 μm d_f.

3.1.2 Liquid Chromatography

Besides the advantages offered by GC, nowadays reversed-phase HPLC is the technique of choice for the separation of polar organic pollutants, silica-bonded columns being preferred [92]. The size parameters of the columns are typically as follows: (1) length in the range 10–25 cm, (2) internal diameter 2.1–4.6 mm and (3) particle sizes 3–5 μm. Gradient elution represents the most common strategy in separation. The mobile phases generally used are acetonitrile, methanol or mixtures of both solvents, obtaining in the latter case shorter retention times and better resolution of the analytes. In order to obtain an efficient retention of the analytes in the column and to improve the sensitivity of MS detection, mobile phase modifiers, buffers and acids are recommended and widely used. The selection of such modifiers strongly depends on the physico-chemical properties of target compounds and their pK_a values. The most common ones include ammonium acetate, ammonium formiate, tri-*n*-butylamine (TrBA), formic acid and acetic acid. Typical concentrations of the salts range from 2 to 20 mM, since it has been observed that higher concentrations could lead to a reduction of the signal intensities [92].

Shortening the analysis times is important for attaining the high sample throughput often required in monitoring studies. This objective can be achieved by shortening the columns and increasing the flow velocity, decreasing the particle size of the stationary phase and finally increasing the temperature, which enhances diffusivity thus allowing working at higher flow rates. These principles are both applied in the Acquity UPLC (ultra-performance liquid chromatography) system, produced by Waters Corporation (Manchester, UK) and in the 1200 Series RRLC (rapid resolution LC) from Agilent Technologies. Both systems use rather short columns (50–100 mm, 4.6 mm ID) packed with sub-2-μm porous particles, allowing very short chromatographic runs. However, the negative effect of using a small particle size is high back-pressure generation (reducing the particle size by a factor of 3 results in an increase in the backpressure by a factor of 27) [92]. Even though the application of UPLC is promising, its application to environmental analysis is still rare. Petrovic et al. [93] developed a UPLC-QqTOF-MS method for screening and confirmation of 29 pharmaceutical compounds belonging to different therapeutic classes in wastewaters, including analgesics and anti-inflammatories, lipid-regulating agents, cholesterol-lowering statin agents, psychiatric drugs, anti-ulcer agents, histamine H_2 receptor antagonists, antibiotics and beta-blockers. UPLC, using columns packed with 1.7-μm particles, enabled elution of target analytes in much narrower, more concentrated bands, resulting in better chromatographic resolution and increased peak height. The typical peak width was 5–10 s at the base, permitting very good separation of all compounds in

10 min, which represented an approximate threefold reduction in the analysis time in comparison to conventional HPLC as shown in Fig. 1.

One of the main problems encountered in quantitative LC analysis and a main source of pitfalls is the existence of matrix effects in general, and the ion suppression phenomenon in particular. The ionization suppression or enhancement may severely influence the sensitivity, linearity, accuracy and precision of quantitative LC analysis. Therefore, any study dealing with analysis of complex samples should include a matrix effect study, and if relevant ion suppression (or signal enhancement) occurs, additional procedures should be applied for correction and/or minimization of inaccurate quantification.

There are several strategies to reduce matrix effects, i.e. selective extraction, effective sample clean-up after the extraction, or improvement of the chromatographic separation. Sometimes, these approaches are not the appropriate solutions because they could lead to analyte losses as well as long analysis times [94]. Recently, several strategies have been adopted as standard practices [95–98]. The most often applied approach consists of the use of suitable calibration, such as external calibration using matrix-matched samples, standard addition or internal standard calibration using structurally similar unlabelled pharmaceuticals or isotopically labelled standards. Other approaches include a decrease of the flow that is delivered to the ESI interface, as well as the dilution of sample extracts. However, the most recommended and versatile approach is isotope dilution, which consists of the use of an isotopically labelled standard for each target compound [99]. But such an approach is expensive and in many cases suffers from a lack of isotopically labelled compounds for all target analytes.

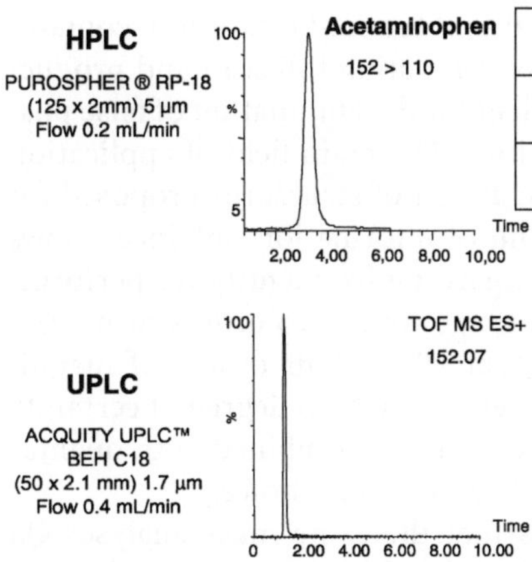

Separation technique	Peak width	Chromatographic run
HPLC	0.5 – 1 min	30 min separation Total run **45 min**
UPLC	5 - 10 s	10 min separation Total run **14 min**

Fig. 1 UPLC versus HPLC chromatograms for the determination of the analgesic acetaminophen (paracetamol) in the PI mode, showing the reduced peak width and increased peak height achieved with UPLC, which results in an improved sensitivity, reduced spectral overlap in complex mixtures and improved MS spectral data

3.2
Detection Systems

The rapid developments in the field of tandem MS/MS have transformed it into a key technique for environmental analysis, replacing other detectors widely used in the past, such as fluorescence and UV detectors for LC and flame ionization (FID), electron capture (ECD) and photoionization (PID) detectors for GC. While tandem MS/MS is mainly coupled to LC, replacing LC-MS due to its higher sensitivity and selectivity, single mass spectrometry is generally attached to GC, mainly using quadrupole, ion trap (IT) and time of flight (TOF) analysers. The latter is mainly applied when working with GC×GC devices.

With regard to LC-MS/MS, triple quadrupole (QqQ) mass analysers have become the most widely used analytical tool in the determination of emerging contaminants in environmental samples. Triple quadrupole instruments gather a variety of scan functions and modes, such as product ion scan, precursor ion scan, neutral loss and multiple reaction monitoring (MRM) mode. LC-MS/MS (QqQ) has been mostly applied to the determination of target analytes, using the selected reaction monitoring (SRM) mode and reaching typically ng L^{-1} detection limits [92].

Although the sensitivity, selectivity and efficiency of the MRM approach are excellent, qualitative information, needed to support the structural elucidation of compounds other than target analytes, is lost [92]. This drawback can be overcome by using the hybrid MS systems, such as QqTOF or QqLIT. The acceptance of QqTOF-MS for environmental analysis in the last few years has been significantly improved and the number of methods reported in the literature is steadily increasing [92].

QqTOF is mainly used as an unequivocal tool for confirmation of contaminants detected. Its unique characteristic of generating full scan and product ion scan spectra with exact masses is excellent for the elimination of false positives and avoiding interpretation ambiguities. The main field of application is the identification of unknowns and elucidation of structures proposed for transformation products, where the amount of information obtained allows secure identification of compounds [92]. Regarding its quantitative performance, QqTOF has a lower linear dynamic range (over two orders of magnitude) with respect to QqQ instruments (typically > four orders of magnitude) [92]. However, when the application requires a high degree of certainty or is aimed at multiple tasks, such as target analysis combined with qualitative investigation of unknowns, its use could be a viable choice.

Regarding QqLIT, its unique feature is that the same mass analyser Q3 can be run in two different modes, retaining the classical triple quadrupole scan functions such as MRM, product ion, neutral loss and precursor ion while providing access to sensitive ion trap experiments [100] (see Fig. 2). This allows very powerful scan combinations when performing information-

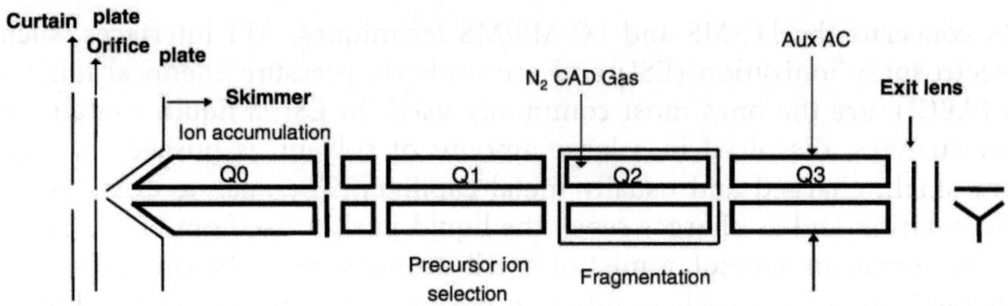

Scan Type	Q1	Q2	Q3
Q1 Scan	Resolving Scan	RF-only	RF-only
Q3 Scan	RF-only	RF-only	Resolving (Scan)
Product Ion Scan (PIS)	Resolving (Fixed)	Fragment	Resolving (Scan)
Precursor Ion Scan (PI)	Resolving (Scan)	Fragment	Resolving (Fixed)
Neutral Loss Scan (NL)	Resolving (Scan)	Fragment	Resolving (Scan Offset)
Selected Reaction Monitoring (SRM)	Resolving (Fixed)	Fragment	Resolving (Fixed)
Enhanced Product Ion Scan (EPI)	Resolving (Fixed)	Fragment	Trap/Scan
MS^3	Resolving (Fixed)	Fragment	Isolation/frag trap/scan
Time delayed frag capture Product Ion (TDF)	Resolving (Fixed)	Trap/No frag	Frag/trap/scan
Enhanced Q3 single MS (EMS)	RF-only	No frag	Trap/Scan
Enhanced Resolution Q3 Single MS (ERMS)	RF-only	No frag	Trap/Scan
Enhanced Multiply Charged	RF-only	No frag	Trap/empty/scan

Fig. 2 Scheme of the QqLIT instrument (QTRAP, Applied Biosystems/Sciex) and description of the various triple quadrupole and trap operation modes

dependent data acquisition. In the case of small molecules, qualitative and quantitative work can be performed concomitantly on the same instrument. The very fast duty cycle of QqLIT provides a superior sensitivity over that of traditional QqQ and ion trap and allows one to record product ion scan spectra for confirmation purposes without compromising signal-to-noise (S/N) ratio. Also the resolution and accuracy are higher and these peculiarities improve the ion selection capability for complex mixtures, i.e. improve the instrumental selectivity. Although environmental applications are still scarce, a few recent papers reported on the application of a hybrid QqLIT for trace level determination of emerging contaminants, such as perfluorinated chemicals, herbicides and pharmaceuticals [92].

3.3
Ionization Sources

For GC-MS instruments, the most common ionization sources employed are electron impact (EI) or chemical ionization, either in negative (NCI) or positive mode (PCI). GC-NCI-MS is mainly used for compounds containing bromine or chlorine ions, such as PBDEs.

As concerns the LC-MS and LC-MS/MS techniques, API interfaces, such as electrospray ionization (ESI) and atmospheric pressure chemical ionization (APCI), are the ones most commonly used. In ESI, a liquid containing target analytes, dissolved in a large amount of solvent, is pushed through a very small, charged and usually metal capillary. The analyte exists as an ion in solution and as charges repel, the liquid pushes itself out of the capillary and forms an aerosol, a mist of small droplets about 10 μm across. An uncharged carrier gas such as nitrogen is sometimes used to help nebulize the liquid and evaporate the neutral solvent in the droplets. As the solvent evaporates, the analyte molecules repel each other and break up the droplets. This process repeats until the analyte is free of solvent and is a lone analyte ion. This process is known as Coulombic fission because it is driven by Coulombic forces between charged molecules. On the other hand, in APCI analytes are already vaporized when introduced into the detector. In this technique, the mobile phase containing eluting analytes is heated to a relatively high temperature (above 400 °C) and sprayed with high flow rates of nitrogen, generating an aerosol cloud which is subjected to a corona discharge to generate analyte ions. These techniques are especially suitable for the determination of low volatility and thermolabile compounds as well as polar substances. ESI is very useful for the analysis of macromolecules because it overcomes the propensity of such molecules to fragment when ionized.

Recently, a new API interface has been developed, the so-called atmospheric pressure photoionization (APPI) interface [101, 102]. APPI is a modification of the APCI source in which the corona is replaced by a gas discharge lamp, emitting radiation in the UV region that is able to selectively ionize the analytes in the presence of the LC mobile phase. Improved performance of APPI can be achieved by adding a dopant, which is a mobile phase additive, like acetone or toluene, which is first ionized itself and then aids ionization of the analytes in further reactions [103]. Compounds like naphthalene, acridine, diphenyl sulphide and 5-fluorouracil could be ionized by an APPI source. Despite being a very new approach, APPI-MS is expected to become an important complementary technique to APCI for low and non-polar analytes in the future [103].

4 Emerging Contaminants

4.1 Fluorinated Alkyl Substances (FASs)

FASs are a group of compounds of anthropogenic origin used in many industrial and consumer products, such as polymers and surfactants. They have been widely used to synthesize products that resist heat, oil, stains, grease and water, due to their unique properties [104].

FASs include the perfluoroalkyl sulphonates (perfluorooctane sulphonate (PFO) and related chemicals, such as *N*-methyl and *N*-ethyl perfluorooctane-sulphonamidoethanol, and also short- and long-chain perfluoro sulphonate acids), the perfluoroalkyl carboxylates (perfluorooctanoate (PFOA) and fluorotelomer alcohols (FTOHs)) and the short- and long-chain perfluoroalkyl acids (e.g. perfluorodecanoic acid (PFDA) [105]). Other substances, such as PFHS and PFBS, considered as "related substances" to PFOs because they have the same moiety ($C_8F_{17}SO_2$ group), are included in the group of PFAs as, once present in the environment, they may decompose to generate PFOs. Many of the degradation products of FASs have been found in the environment throughout the world, but PFOs and PFOA are the two most widely detected groups. Because of the strong carbon–fluorine (C–F) bond associated with their chemical structure, they are environmentally persistent substances and have been detected in human blood, water, soils, sediments, air and biota [105].

Due to their high production worldwide, in October 2000 the US EPA proposed a significant new use rule (SNUR) for 88 PFO-related substances [105]. On the other hand, PFOs and related substances have also been on the agenda of the Organization for Economic Co-operation and Development (OECD) since the year 2000 [105]. In the EU, there is currently no legislation on their use associated with their potential environmental and/or human health effects. However, some legislation which generally applies to the release of substances to the environment may be relevant to the release of PFOs. Therefore, the IPPC Directive 96/61/EC includes fluorine and its compounds in the "indicative list of the main polluting substances to be taken into account if they are relevant for fixing emission limit values". There are several reviews devoted to their analysis in environmental samples [105, 106]. However, these compounds present several difficulties during their analysis, as indicated in the section below.

4.1.1
Background Contamination Problems

The analysis of PFAs is rather difficult due to several background contamination problems not only coming from the materials used for sample collection and preparation, but also from the instrumental techniques [104, 107–109]. Therefore, one source of experimental contamination is the use of materials made of, or containing, fluoropolymers, such as polytetrafluoroethylene (PTFE) or perfluoroalkoxy compounds, which should be avoided. Taniyasu et al. [107] performed several experiments to assess possible sources of contamination, from sample collection materials to solvents used. They found that polypropylene sample bottles used for sample collection and storage contained PFOA. In the evaluation of two widely employed SPE cartridges, the Oasis hydrophilic–lipophilic balanced (HLB) and Sep-Pak C_{18}, considerable amounts of PFOA, PFOs, PFHS and PFBS were detected, the latter being the one showing higher concentrations. Even purified water was found to be another possible source of contamination. In the light of these concerns, water samples are collected in polyethylene or polypropylene bottles rinsed with methanol and deionized water prior to use. Glass is avoided because analytes tend to bind it and some authors centrifuge water samples, as an alternative to filtration, to avoid possible adsorption of PFOs onto the filter and subsequent loss of analyte [110].

Moreover, during instrumental analysis, especially when working with LC-MS or tandem MS/MS detection, significant instrumental contamination problems can occur. Yamashita et al. [109] determined that the HPLC tubing, internal fluoropolymer parts and autosampler vial septum were potential sources of PFA contamination during LC analysis. Therefore, it is recommended to replace the PTFE HPLC tubing with stainless steel and polyetheretherketone (PEEK). Moreover, the same authors isolated the degasser and solvent selection valves, which contain fluoropolymer coatings and seals from the HPLC system, and the solvent inlet filters were replaced by stainless steel ones. Finally, autosampler vial caps made of Viton fluoropolymers or polyethylene were used, as they reduced considerably the instrumental blank concentrations.

4.1.2 Sample Preparation

Fluorinated alkyl substances have been mainly analysed in biological samples and environmental waters [105]. Concerning their determination in aqueous matrices, liquid–liquid extraction (LLE) and solid-phase extraction (SPE) are the traditional methods used for enrichment and isolation of target analytes, mainly using Oasis HLB, octadecyl C_{18} bonded silica and Oasis WAX adsorbents (see Table 1) [105]. On-line direct analysis using diverse preconcentration columns has been proposed by several authors [18, 106, 111–113], to speed up sample preparation.

Only Higgins et al. [114] have determined the presence of fluorinated compounds in sediments. Extraction was performed using a heating sonication bath and afterwards a clean-up procedure with C_{18} SPE cartridges. These compounds have also been determined in sludges by Higgins et al. [114] and Schröder et al. [115]. The former applied the same treatment as for the sediments. The latter compared the efficiency of three extraction techniques (Soxhlet, hot vapour and PLE), PLE being the one yielding better performances. After extraction, crude extracts are purified, generally using SPE with C_{18} cartridges (see Table 2).

Table 1 Representative methods, indicating the extraction and detection techniques, for the determination of the selected groups of emerging contaminants in environmental waters

Compounds	Matrix	Extraction method	Purification or derivatization for GC	Detection	GC/LC column	LC mobile phase	LOD (ng/L)	Refs.
MTBE, degradation products and other gasoline additives	Influent/ effluent wastewaters	P&T	–	GC-EI-MS				[362]
	Influent/ effluent wastewaters	HS-SPME	–	GC-EI-MS				[351]
	Ground water	P&T with Tenax® silica gel–charcoal at room temperature. Desorption with He at 225 °C	–	GC-EI-MS	Capillary fused silica DB-624 (75 m × 0.53 mm)		1–110	[347]
PFOs	Surface water	SPE (Presep-C cartridges)	–	LC-ESI-MS	Zorbax XDB C_{18} (2.1 × 150 mm)	AcN-H_2O (10 mM NH_4Ac)	0.04–0.1	[111, 112]
PFOs, *N*-EtFOSAA	Wastewater	SPE (Waters, Oasis HLB 1 g)	–	LC-ESI-MS/MS	Zorbax SB C_8 (3.0 × 150 mm)	A: MeOH/AcN (50%) 0.15% HOAc B: Water 0.15% HOAc	0.06–0.1	[363]

Table 1 (continued)

Compounds	Matrix	Extraction method	Purification or derivatization for GC	Detection	GC/LC column	LC mobile phase	LOD (ng/L)	Refs.
PFNA PFOSA FTOHS	Seawater	SPE (Oasis WAX)	–	LC-ESI-MS/MS	Guard column: XDB-C_8 (2.1×12.5 mm) Column: Betasil-C_{18} (2.1×150 mm)	A: H_2O (2 mM NH_4Ac) B: MeOH	1.8 pg/L 1pg/L 0.01–1	[107]
E1, E2, 17α-E2, EE	Surface water Drinking water STP effluent	SPE (Lichrolut EN)	Derivatization with 10% PFBCl in toluene	GC-NCI-MS	DB5MS (60m×0.32 mm, 0.25 μm)	–	0.05–0.15	[185]
E1, E2, E3, EE	Ground water	SPE (Oasis HLB)	Derivatization with PFBBR + TMSI (LLE with water and hexane)	GC-NCI-MS/MS	DB5-XLB (60m×0.25 mm, 0.25 μm)	–	0.2–0.6	[134]
E1, E2, EE	Drinking, ground, surface and wastewater	SPE (Bakerbond C_{18})	For WWTP influent SPE (silica gel)	LC-ESI (NI) MS/MS	RP-C_8 Hypersil MO5 (100×2.1 mm, 5 μm)	A: ACN/MeOH B: H_2O	0.1–2	[167, 168]
E1, E2, E3, EE, DES, E2-17G, E1-3S, E2-17 Acet.	Ground, river and treated waters	Fully automated on-line SPE (PLRP-s)	–	LC-ESI (NI) MS/MS	Purospher STAR-RP18e (125×2 mm, 5 μm Merck)	A: ACN B: H_2O	0.01–0.38	[138]

Table 1 (continued)

Compounds	Matrix	Extraction method	Purification or derivatization for GC	Detection	GC/LC column	LC mobile phase	LOD (ng/L)	Refs.
E1, E2, E3 + PROG + six androgens	Ground and river water	SPE (Carbograph)	–	LC-APCI (PI) MS/MS	Alltima C_{18} (250×4.6 mm, 5 μm Alltech)	A: ACN B: H_2O 5 mM NH_4Ac	0.5–1	[364]
Antibiotics, β-blockers, psychiatric drugs, anti-inflammatories	Hospital effluent wastewaters	pH adjustment (pH 7) SPE (Oasis HLB)	–	LC-ESI (NI) and (PI) MS/MS	Purospher STAR-RP18e (125×2 mm, 5 μm Merck)	**ESI(+)** A: ACN B: Aq-Formic acid **ESI(–)** A: ACN B: H_2O	4–47	[200]
Anti-inflammatories, lipid regulators, anti-epileptic, β-blockers, antibiotics and other contaminants	River and wastewaters	Natural water pH SPE Oasis HLB	–	LC-ESI (NI) and (PI) MS/MS	Purospher STAR-RP18e (125×2 mm, 5 μm Merck)	**ESI(+)** A: ACN/MeOH (2:1) B: NH_4Ac 5 m/ HAc **ESI(–)** A: MeOH B: H_2O	0.5–47 RW 1–60 WW	[2]
Analgesics/ anti-inflammatories, lipid regulators, β-blockers, antibiotics, anti-epileptics	Surface water	Sample acidified at pH = 3 SPE Oasis MCX	–	LC-ESI (NI) and (PI) MS/MS		**ESI(+)** and **ESI(–)** A: MeOH B: 2 mM NH_4Ac	5–25	[365]

Table 1 (continued)

Compounds	Matrix	Extraction method	Purification or derivatization for GC	Detection	GC/LC column	LC mobile phase	LOD (ng/L)	Refs.
Tetracycline and sulphonamide antibiotics	Wastewaters	Addition of Na_2EDTA and citric acid (pH<3) SPE Oasis HLB	–	LC-ESI (PI) MS/MS		**ESI(+)** A: AcN B: 0.1% formic acid	30–70	[366]
All musk (no metabolites)	Wastewaters	LLE with hexane SEC (Bio Beads SX-3)	Silica purification	GC/EI-MS	VR-5MS (30 m×0.25 mm, 0.25 μm)		NR	[258]
HHCB, AHTN, ATII, ADBI, AHMI, DPMI, MX, MK	WWTP effluent and surface water	SLLE with pentane, DCM, DCM (at pH 2) Dried with sodium sulphate	–	GC/EI-MS	BPX-5 (30 m×0.25 mm, 0.25 μm)		NR	[234, 235]
HHCB, AHTN	Ground water	SPE (C_{18}) Eluent: acetone/hexane (3:17 VR)	Silica purification	GC/EI-MS	XTI-5 (30 m×0.25 mm, 0.25 μm)		NR	[197]
BDE-15, BDE-28, BDE-47, BDE-100, BDE-99, BD-154, BDE-153, BDE-183	Tap and river water	HF-MMLLE using *n*-undecane as solvent. Extraction time: 60 min; stirring rate: 1200 rpm	–	GC/EI-MS	HP-5 ms (30 m×0.25 mm, 0.25 μm)		0.2–0.9	[320]
BDE-47, BDE-100, BDE-99, BDE-85, BDE-154, BDE-153	River, sea and wastewater	SPME using polydimethylsiloxane (PDMS) rods	–	GC-ECD-MS	HP-5 (30 m×0.32 mm, 0.25 μm)		0.3–5	[367]

Table 1 (continued)

Compounds	Matrix	Extraction method	Purification or derivatization for GC	Detection	GC/LC column	LC mobile phase	LOD (ng/L)	Refs.
α,β,γ-HBCD	Landfill leachate	LLE using DCM SPE Abselut Nexus	–	LC-ESI-MS/MS	Develosil C30-UG-5 (150 mm×2 mm)	**ESI(–)** A: ACN B: H_2O	NR	[368]
APEO, APEC, AP, halogenated derivatives	Surface drinking, and wastewaters	SPE C_{18}	–	LC-ESI (NI)/APCI-MS	Lichrospher RP-18 100 (250×4 mm, 5 μm)	**ESI(–)** A: MeOH B: H_2O **APCI** A: MeOH/ACN (1:1) B: H_2O	5–20 μg for river sediment 5–25 μm for sewage sludge	[277]
AEO, NPEO, CDEA, LAS, NPEGNP, OP	Coastal waters	SPE Lichrolut C_{18}	–	LC-ESI (NI)/APCI-MS	Lichrospher RP-18 100 (250×4 mm, 5 μm)	**AEO, NPEO, CDEA APCI** A: MeOH/ACN (1:1) B: H_2O **LAS, NPEC, NP, OP** **ESI(–)** A: MeOH; B: H_2O	10–150	[279]

Table 2 Representative methods for the determination of the selected groups of emerging contaminants in solid samples, indicating the extraction, purification procedures and detection systems

Compounds	Matrix	Extraction method	Purification or derivatization for GC	Detection	GC/LC column	LC mobile phase	LOD	Refs.
MTBE, degradation products and other gasoline additives	Soil	P&T with Tenax® silica gel–charcoal at room temperature. Desorption with He at 225 °C	–	GC-EI-MS	Capillary fused silica DB-624 (75 m×0.53 mm)		0.01–1.44 μ/kg	[350]
PFOs	Sediments	3 extractions with 90:10 (v/v) MeOH and 1% HOAc	SPE C_{18}	LC-ESI-MS/MS	Targa Sprite C_{18} (40×2.1 mm)	MeOH-H_2O 2 mM NH_4Ac	0.04–0.07 ng/L 0.109 ng/g	[114]
PFOA, PFHS, *N*-MeFO, SAA, *N*-EtFOSAA, anionic, non-ionic	Sewage sludge	PLE [EtOAc/DMF (8:2), MeOH/H_3PO_4 (95:5), MeOH/H_3PO_4 (99:1), MeOH/H_3PO_4 (99:1)] 150 °C, 10 714 kPa	–	LC-ESI-MS	PF-C_8 column (150×4.6 mm) filled with spherical perfluorinated RP-C_8 material (5 μm)	A: MeOH B: MeOH/H_2O (80:20) (2 mM diethyl ammonium)	0.6 ng/g	[115]
E1, E2, α-E2, E3, MES (+BPA, NP)	River sediment	Ultrasonication (acetone/DCM, 1:1)	LLE with DCM + silica gel fractionation. Derivatization: PFPA	GC-EI-MS	HP-5MS (30 m×0.25 mm, 0.25 μm)		0.6–2.5 ng/g	[151]

Table 2 (continued)

Compounds	Matrix	Extraction method	Purification or derivatization for GC	Detection	GC/LC column	LC mobile phase	LOD	Refs.
E1, E2, EE, MES	Sludge	Ultrasonication (MeOH + acetone)	GPC Biobeads SX-3 SPE (silica gel) Derivatization: MSTFA/TMSI/ DTE (1000:2:2, v/v/w)	GC-(IT)-MS/MS	XTI-5 (30 m×0.25 mm, 0.25 μm)		2–4 ng/g	[149]
17G, E2–3, 17diS E1, E2	Estuary sediment	Sonication (MeOH)	SPE (Lichrolut EN + BondElut C_{18}) + NP-LC fractionation	LC-ESI (NI)-TOF-MS	Betasil C18 (150×2.1 mm, 3 μm, Keystone Scientific)	A: AcN B: H_2O	0.03–0.04 ng/g	[152]
E1, E2, E3, EE, DES (+ progestins)	River sediment	Sonication (acetone: methanol, 1:1)	SPE (C_{18})	LC-ESI (NI)-MS	Lichrospher 100 RP-18 (250×4 mm, 3 μm, Merck)	A: AcN B: H_2O	1–2 ng/g	[153]
Tetracycline, macrolide and sulphonamide antibiotics	Agricultural soils	PLE MeOH/citric acid (1:1, v/v) adjusted to pH = 4.7 with NaOH	Dilute PLE extracts to MeOH content < 10%. Purification with SAX-Oasis HLB in tandem	LC-ESI (PI)-MS/MS	X-terra MS-C_{18} (100×2.1 mm, 3.5 μm, Merck)	A: MeOH B: Aq. formic acid	8–22 μg/L	[194]

Table 2 (continued)

Compounds	Matrix	Extraction method	Purification or derivatization for GC	Detection	GC/LC column	LC mobile phase	LOD	Refs.
Tetracycline, sulphonamides, fluoro-quinolone antibiotics and trimethoprim	Arable soils fertilized with manure	**TCs, SAs and TMP** MeOH/EDTA-McIlvaine buffer pH = 6 (90:10, v/v) **FQs** AcN acidified with 2% HCOOH	**TCs, SAs and TMP** SPE C_{18} **FQs** LLE with hexane	**TCs, SAs and TMP** LC-ESI (PI) MS/MS **FQs** LC-ESI (PI) MS	**TCs, SAs and TMP** Luna (Phenomenex) C_8 (150×2 mm, 5 μm) **FQs** Luna (Phenomenex) C_8 (150×3 mm, 5 μm)	**TCs, SAs and TMP** A: ACN B: H_2O C: 0.5% HCOOH 10 mM NH_4OAc **FQs** A: ACN 0.01% HCOOH B: H_2O 0.01% HCOOH	1.6–18 (ng/mL)	[369]
Analgesics and anti-inflammatories, lipid regulators, antibiotics and ivermectin	River sediment	Ultrasound **Acidic compounds** Acetone/HAc (20:1, v/v) + ethyl acetate **Antibiotics** MeOH/acetone + ethyl acetate	Dilute extracts **Acidic compounds** Acidify at pH = 2 SPE Oasis MCX **Antibiotics** Acidify at pH = 3 SPE Lichrolut EN + C_{18} **Ivermectin**	**Acidic compounds** LC-ESI (NI) MS/MS **Antibiotics** LC-ESI (PI) MS	All compounds Lichrospher RP-18 (125×3 mm, 5 μm, Merck)	**Acidic compounds** A: ACN B: H_2O pH = 2.9 (with HAc) **Antibiotics** A: Eluent B + AcN B: 20 mM NH_3 at pH = 5.7 with HAc	**Acidic compounds** 0.4–20 ng/g **Antibiotics** 3–20 ng/g	[195]

Table 2 (continued)

Compounds	Matrix	Extraction method	Purification or derivatization for GC	Detection	GC/LC column	LC mobile phase	LOD	Refs.
			Add NH_4Ac buffer SPE Lich-rolut EN			**Ivermectin** A: ACN 10% B B: 15 mM NH_4 AC + HAc (pH = 4)		
All musks and metabolites (except DPMI)	Activated sludge	LLE with hexane	Silica purification	GC-MS/MS GC-EI-MS	DB-1 (60 m×0.25 mm, 0.25 μm)		NR	[265, 370]
HHCB, AHTN, ATII, ADBI, AHMI, DPMI, MX, MK, MA, MM, MT	Digested sludge	Dried with sodium sulphate Soxhlet extraction with DCM Sulphur removed with copper in flask during extraction	Silica/alumina purification (layered) SEC (Bio Beads S-X3) Silica/alumina purification	GC-EI-MS	HP-5MS (30 m×0.25 mm)		NR	[261]
All musks	Sludge	SFE with acetone/DCM (1:1)	Silica purification Sulphur removed with copper	GC-NCI/MS GC-EI-MS	HP-5MS (30 m×0.25 mm, 0.25 μm)		NR	[371]

Table 2 (continued)

Compounds	Matrix	Extraction method	Purification or derivatization for GC	Detection	GC/LC column	LC mobile phase	LOD	Refs.
Mono-hepta-BDEs (39 compounds)	Marine and river sediment	PLE (Cu + Al_2O_3 1:2) using DCM:C6 (1:1) as solvent	–	GC-NCI-MS	HP-5MS (30 m×0.25 mm, 0.25 μm)		1–46 pg/g	[326]
α, β, γ-HBCD	Sediments	Soxhlet (acetone:C6, 3:1)	LLE with H_2SO_4 + GP + SiO_2	LC-ESI (NI) MS	Luna C_{18} (150×2 mm, 5 μm, Merck)	A: AcN + 10 mM NH_4OAc B: H_2O + 10 mM NH_4OAc	NR	[372]
Di-hexa BDEs + deca-BDEs (14 compounds)	Sewage sludge	PLE (DCM:C6, 1:1)	H_2SO_4 + SiO_2 H_2SO_4 + Al_2O_3	**Di-hexa BDE:** GC-MS/MS **Deca-BDE:** GC-NCI-MS	NR		NR	[373]
Mono-deca BDEs (40 compounds), total HBCD	Fish tissue	PLE (Al_2O_3, DCM:C6, 1:1)	–	GC-NCI-MS	HP-5MS (30 m×0.25 mm, 0.25 μm)		2–19 pg/g (wet-weight)	[306]
Tri-deca BDEs (27 compounds)	Fish tissue	PLE (DCM)	GPC + SiO_2	GC-NCI-MS	NR		NR	[374]
Non-ionic surfactants, NPEO, AEO, CDEA	Sewage sludge	Sonication (DCM/MeOH, 3:7)	SPE C_{18}	LC-ESI (NI)/APCI-MS	Lichrospher RP-18 100 (250×4 mm, 5 μm)	**ESI** (–) A: MeOH B: H_2O **APCI** A: ACN B: H_2O	5–25 μg/kg	[277]

Table 2 (continued)

Compounds	Matrix	Extraction method	Purification or derivatization for GC	Detection	GC/LC column	LC mobile phase	LOD	Refs.
APEO, APEC, AP, halogenated derivatives	River sediment, sludge	Sonication (DCM/MeOH, 3:7)	SPE C_{18}	LC-ESI (NI)/APCI-MS	Lichrospher RP-18 100 (250×4 mm, 5 μm)	**ESI(–)** A: MeOH B: H_2O **APCI** A: MeOH/ ACN (1:1) B: H_2O	20–100 μ/kg	[277]
Ionic surfactants LAS, SPC	Marine sediment	Soxhlet (MeOH)	SPE C_{18}	LC-FL	Lichrosorb RP-18 (250×4.6 mm, 10 μm)	A: MeOH/H_2O (80:20) with 1.25 mM tetraethyl-ammonium B: H_2O	5–10 μ/kg	[375]

4.1.3 Instrumental Analysis

Fluorinated surfactants can be detected by ^{19}F NMR, gas and liquid chromatography–mass spectrometry and liquid chromatography coupled to tandem mass spectrometry [105], the latter two being the most widely employed.

^{19}F NMR spectroscopy is a non-specific method, as it determines the presence of CF_2 and CF_3 moieties [116, 117]. Moody et al. [117] compared the results achieved by this technique with LC-MS/MS, showing discrepancies between the two methods. With ^{19}F NMR the total content of perfluorinated compounds was higher than that calculated by LC-MS/MS, attributed to the presence of other surfactants in the samples which yielded a similar ^{19}F NMR spectrum to perfluoroalkanesulphonates and perfluorocarboxylates [105].

Gas chromatography–mass spectrometry can be used for the direct determination of neutral and volatile FASs, such as sulphonamides or fluorotelomer alcohols, which have high vapour pressures [105]. Perfluorocarboxylates have been quantitatively determined by GC-MS after derivatization of the carboxylates to their methyl esters [116, 117]. However, PFOs was not able to be detected by such a method [117]. Although perfluoroalkane sulphonate esters may be formed during the derivatization step, the esters are unstable because of the excellent leaving group properties of perfluoroalkane sulphonates [105]. Thus, despite the fact that some fluorinated surfactants can be analysed by GC-MS, this technique is not so useful for multi-residue analysis of all groups of PFAs [105]. The drawbacks offered by both ^{19}F NMR and GC-MS and the multiple advantages presented by LC-MS and LC-MS/MS, in terms of sensitivity and selectivity, have made these techniques the preferred tools for the instrumental analysis of PFAs in environmental samples. Other detectors coupled to LC include fluorescence detection for the determination of perfluorocarboxylic acids [118], ion-exclusion chromatography with conductimetric detection for perfluorocarboxylic acid and perfluorosulphonates [119, 120] and LC with conductimetric detection for perfluorosulphonates [121].

Electrospray ionization (ESI) working in the negative ion (NI) mode is the interface most widely used for the determination of anionic perfluorinated surfactants. APCI is not suitable for the determination of PFOs due to their ionic nature. The ESI interface has also been optimized for the determination of neutral compounds, such as the sulphonamides PFOSA, Et-PFOSA and *t*-Bu-PFOs [122]. Takino et al. [110] developed a method based on an APPI interface, which would alleviate matrix effects found with ESI interfaces.

Chromatographic separation of fluorinated compounds has been mainly carried out using both RP-C_{18} and RP-C_8 materials. However, RP-C_{18} presented some interferences, enhancing analyte signals and, therefore, the

Fig. 3 LC-ESI(NI)-MS chromatograms obtained in the SIM mode for a standard solution containing **a** perfluorocarboxylic acids and **b** sulphonates and neutral FASs. Reprinted with permission from [376] ▸

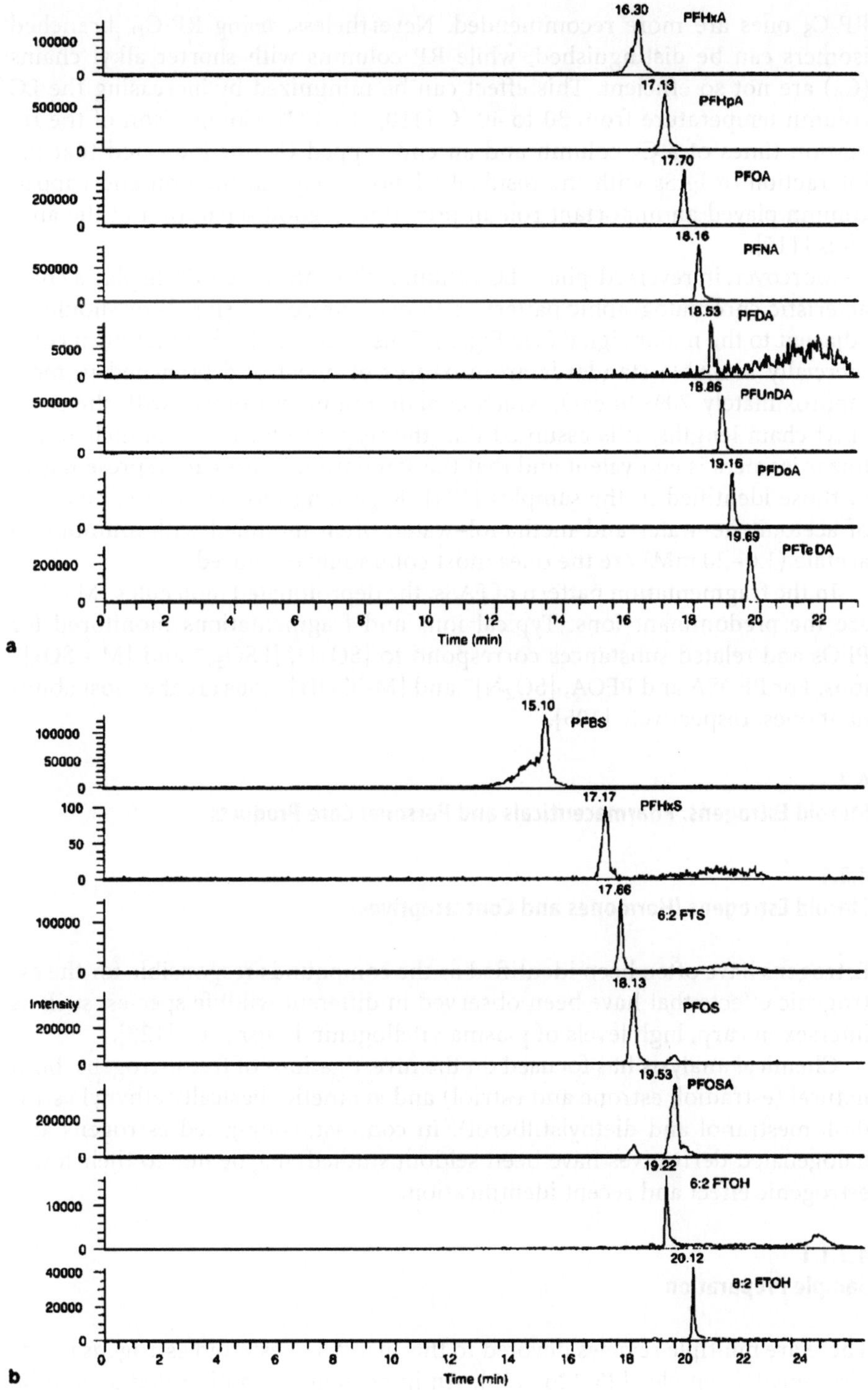
16.30 PFHxA
17.13 PFHpA
17.70 PFOA
18.16 PFNA
18.53 PFDA
18.86 PFUnDA
19.16 PFDoA
19.69 PFTeDA
Time (min)
a
15.10 PFBS
17.17 PFHxS
17.66 6:2 FTS
18.13 PFOS
Intensity
19.53 PFOSA
19.22 6:2 FTOH
20.12 8:2 FTOH
Time (min)
b

RP-C_8 ones are more recommended. Nevertheless, using RP-C_{18} branched isomers can be distinguished, while RP columns with shorter alkyl chains (C_8) are not so efficient. This effect can be minimized by increasing the LC column temperature from 30 to 40 °C [110, 112, 123]. Comparison of the retention times of a C_8 column and an end-capped C_8 one indicated that the interaction of FASs with the residual silanol groups in the non-end-capped column played an important role in providing a good separation of the analytes [115].

Moreover, in reversed-phase LC columns, the FAS standards display a characteristic chromatographic pattern with two unresolved signals or shoulders adjacent to the major signal (see Fig. 3). This is due to the fact that most commercially available standards are mixtures of linear and branched isomers (approximately 70% linear), which contain impurity isomers with the same alkyl chain lengths. It is assumed that the response factor for branched and linear isomers is equivalent and that the standard mixtures are representative of those identified in the samples [124]. Regarding mobile phases, mixtures of acetonitrile–water and methanol–water, often modified with ammonium acetate (1.0–20 mM) are the ones most commonly employed.

In the fragmentation pattern of FASs, the deprotonated molecules $[M-H]^-$ are the predominant ions. Typical ions and fragmentations monitored for PFOs and related substances correspond to $[SO_3]^-$, $[FSO_3]^-$ and $[M-SO_3]^-$ ions. For PFOSA and PFOA, $[SO_2N]^-$ and $[MCOOH]^-$ ions are the most abundant ones, respectively [105].

4.2 Steroid Estrogens, Pharmaceuticals and Personal Care Products

4.2.1 Steroid Estrogens (Hormones and Contraceptives)

Estrogens have often been identified as the compounds responsible for the estrogenic effects that have been observed in different wildlife species, such as intersex in carp, high levels of plasma vitellogenin in fish, etc. [125].

Chemical analysis has focused on the investigations of free estrogens, both natural (estradiol, estrone and estriol) and synthetic (basically ethynyl estradiol, mestranol and diethylstilberol). In contrast, conjugated estrogens and halogenated derivatives have been seldom studied, maybe due to their lower estrogenic effect and recent identification.

4.2.1.1 Sample Preparation

There are multiple reviews devoted to the analysis of esteroid estrogens in environmental samples [25, 126–133]. An important precaution that should be taken into account when analysing steroid estrogens in tap water, or water samples that could contain chlorine, is the addition of sodium thiosulphate immediately after collection in order to avoid losses of target analytes [134].

Extraction of estrogens from water samples has usually been carried out by off-line SPE using either disks or, most frequently, cartridges (see Table 1), with octadecyl C_{18}-bonded silica, polymeric graphitized carbon black (GCB) and Oasis HLB being the most widely employed cartridges [134–136]. On the other hand, many works are based on the use of on-line SPE [129, 137, 138], using the same extraction materials as indicated for off-line SPE. To elute compounds trapped in the SPE cartridges, methanol is the solvent generally used. However, Isobe et al. [136] determined that adding 5 mM of TEA to 10 mL of methanolic solution, as an ion pair reagent, improved the efficiency of elution, thus achieving higher recoveries for conjugates which were not effectively removed by only using methanol.

Other widely employed materials to isolate steroid estrogens from water samples are molecularly imprinted polymers (MIPs) [25, 38, 139]. Some recent works have also proposed the use of SPME, using fibre and in-tube SPME, in combination with either LC or GC instruments [140, 141, 143].

As concerns the determination of esteroid estrogens in solid samples, the analytical methods are generally adapted from those developed for water samples, incorporating additional purification steps of crude extracts prior to instrumental analysis [144]. Extraction techniques more commonly used are pressurized liquid extraction (PLE) [145, 146], microwave-assisted extraction (MAE) [147] and, more frequently, ultrasonication [148–153], using methanol [148, 152], methanol/acetone [145, 146, 149, 153], acetone/dichloromethane [151], ethyl acetate [154, 155] or dichloromethane/water [150] as extraction solvents. Some of the most representative methods are summarized in Table 2.

Purification of extracts is generally carried out by liquid–liquid extraction (LLE) [156–158], HPLC fractionation [156, 159–162], gel permeation chromatography (GPC) [158], immunoaffinity (IA) extraction [25] or SPE using Florisil [136, 157], C_{18} sorbents [132, 156, 159, 160], silica gel [163–169] and restricted access materials (RAMs).

4.2.1.2
Instrumental Analysis

In the past, the techniques most commonly used for the environmental analysis of estrogens have been immunoassays and, to a greater extent, GC-MS. The former are simple and sensitive but they can have false positive results due to the influence of coexisting materials present in the sample matrix. On the other hand, GC-MS and GC-MS/MS are also highly sensitive methods, but derivatization is required prior to analysis [141]. Moreover, these methodologies are mainly based on the determination of unconjugated (i.e. free) estrogens, unless intermediate hydrolysis steps are performed [136, 170]. LC-MS and especially LC-MS/MS are the preferred tools nowadays [171, 172], which allow the determination of both conjugated and free estrogens without derivatization and hydrolysis.

Enzyme-linked immunosorbent assay (ELISA) and radioimmunoassay (RIA) are by far the most common *bioassays* used for the determination of estrogens. Several recent works have reported their application in the analysis of estrogens in environmental matrices, such as water [173–176], sludge and manure, although they have been more extensively used for the analysis of biological samples in clinical studies. Their main advantages are ease of use, relatively simple protocol and fairly good sensitivity. Bioassays are also used to measure the estrogenic (endocrine disrupting) activity of sample extracts or of chemicals. The in vitro and in vivo assays available for this purpose have been recently reviewed [177, 178]. Many bioassays show potential for development as biosensors [179, 180].

On the other hand, GC separation has been performed with a variety of capillary columns (DB5-MS, XTI-5, HP Ultra II, etc.), using helium as carrier gas. Both conventional MS and MS/MS detection have been accomplished in most instances in the electron impact (EI) mode at 70 eV. The use of negative ion chemical ionization (NICI) has been reported on fewer occasions [134, 165, 181–184]. However, it has been observed that the highest sensitivity for the GC-NICI-MS methods is obtained when estrogens have pentafluorobenzyl (PFB) [181, 182], pentafluorobenzoyl [184, 185] and other fluorine-containing derivatives.

Derivatization is generally carried out in the – OH groups of the steroid ring, performed by silylation with reagents such as *N*,*O*-bis(trimethylsilyl)-acetamide (BSA), *N*-methyl-*N*-trimethylsilyltrifluoroacetamide (MSTFA), *N*,*O*-bis(trimethylsilyl)-trifluoroacetamide (BSTFA), or *N*-(*tert*-butyldimethylsilyl)-*N*-methyltrifluoroacetamide (MTBSTFA), which lead to the formation of trimethylsilyl (TMS) and *tert*-butyldimethylsilyl (TBS) derivatives [186]. Some authors reported breakdown of some TMS derivatives with various solvent–reagent combinations, pyridine and dimethylformamide being the most suitable ones [186–188].

LC has been performed by octadecyl silica stationary phases. As mobile phases, mixtures of water/methanol and, more frequently, water/acetonitrile have normally been used, sometimes with added modifiers such as 0.1% acetic acid, 0.2% formic acid or 20 mM ammonium acetate. The interfaces most widely employed are electrospray ionization (ESI) in the negative ion (NI) mode and, to a lesser extent, atmospheric pressure chemical ionization (APCI) in the positive ionization (PI) mode. These API interfaces have been applied in a variety of MS analysers, including quadrupole, ion-trap, orthogonal-acceleration time-of-flight (oaTOF), and combinations of them. Single and triple quadrupole analysers have been the most widely used for the analysis of estrogens, the latter being preferred nowadays. Some works

Table 3 MRM transitions monitored for the determination of steroid estrogens and pharmaceuticals in environmental samples using LC-ESI-MS/MS (QqQ) instruments

Group of substances	Compound	MRM 1	MRM 2
Steroid estrogens	Estriol	287>171 Loss of $C_6H_{12}O_2$	287>145 Loss of $C_8H_{14}O_2$
	Estradiol	287>145 Loss of $C_8H_{14}O$	281>183 Loss of $C_5H_{12}O$
	Estrone	269>145 Loss of $C_8H_{12}O$	269>143 Loss of $C_8H_{14}O$
	Ethynyl estradiol	295>145 Loss of $C_9H_{12}O$	295>159 Loss of $C_{10}H_{14}O$
Anti-inflammatory/ analgesic/antiphlogistic	Ibuprofen	205>161 Loss of CO_2	–
	Ketoprofen	253>209 $[M-H-CO_2]^-$	253>197
	Naproxen	229>185 $[M-H-CO_2]^-$	229>170 $[M-H-C_3H_2O_2]^-$
	Indomethacin	356>312 $[M-H-CO_2]^-$	356>297 $[M-H-C_3H_2O_2]^-$
	Diclofenac	294>250 $[M+H-H_2O]^+$	294>214
	Acetaminophen	152>110 Loss of CH_2CO 150>107 Loss of $COCH_3$	152>93 –
	Fenoprofen	241>197	241>93
	Mefenamic acid	240>196 Loss of CO_2	240>180 $[M-H-CO_2-CH_3]^-$
	Propyphenazone	231>189 $[M-C_3H_7+H]^+$	231>201
	Phenylbutazone	309>160 $[M-(C_6H_5-N-(C_4H_9))]^+$ 362>276	309>181 362>316
Lipid regulating agents	Bezafibrate	360>274 Loss of $C_4H_6O_2$	360>154 Loss of $C_{12}H_{14}O_3$
	Clofibric acid	213>127 $[C_6O_4ClO]^-$	213>85
	Gemfibrozil	249>121 $[M-H-C_7H_{12}O_2]^-$	–
Psychiatric drugs	Carbamazepine	237>194 Loss of HNCO	237>192
	Fluoxetine	310>44 $[M-F_3C_7H_4OC_8H_8]^+$	310>148 $[M-F_3C_7H_4O]^+$
	Paroxetine	330>192 $[M-C_7H_5NO_3]^+$	330>123 $[M-C_{12}H_4NOF]^+$
	Diazepam	285>257 $[M-CO+H]^+$	285>154

Table 3 (continued)

Group of substances	Compound	MRM 1	MRM 2
Macrolide antibiotics	Erythromycin-H_2O	716>522 $[M\text{-}DS\text{-}2H_2O+H]^+$	716>558 $[M\text{-}DS\text{-}H_2O+H]^+$
	Clarythromycin	750>116 $[CL\text{-}OCH_3+H]^+$	750>592 $[M\text{-}DS+H]^+$
	Roxythromycin	838>158 $[DS+H]^+$	838>680 $[M\text{-}DS+H]^+$
	Oleandomycin	689>545 $[M\text{-}oleandrose+H]^+$	689>158 $[DS+H]^+$
	Tylosin	916>723 $[M\text{-}MY+H]^+$	916>174 $[DS\text{-}O\text{-}MY+H]^+$
Tetracycline antibiotics	Chlortetracycline	479>444	479>462
	Doxycycline	445>428	445>410
	Oxytetracycline	461>426	461>443
	Tetracylcline	445>410 $[M\text{-}H_2O\text{-}NH_3+H]^+$	445>427 $[M\text{-}H_2O+H]^+$
Quinolone antibiotics	Ciprofloxacin	332>314 $[M\text{-}H_2O+H]^+$	332>288 $[M\text{-}H_2O\text{-}CO_2+H]^+$
	Ofloxacin	362>344 $[M\text{-}H_2O+H]^+$	–
	Norfloxacin	320>302 $[M\text{-}H_2O+H]^+$	320>302 $[M\text{-}CO_2+H]^+$
	Enrofloxacin	360>342 $[M\text{-}H_2O+H]^+$	360>316 $[M\text{-}CO_2+H]^+$
Sulphonamide antibiotics	Sulphamethoxazole	254>156 $[H_2NPhSO_2]^+$	254>92 $[H_2NPhO]^+$
	Sulphamethazine	279>186 $[M\text{-}H_2NPh]^+$	279>124 [aminodimethyl-pyridine+H]$^+$
	Sulphadiazine	251>156 $[H_2NPhSO_2]^+$	251>108 $[H_2NPhO]^+$
Penicillins	Dicloxacillin	487>160 $[F1+H]^+$	487>311 $[F2+H]^+$
	Nafcillin	432>171 [ethoxynaphthyl]$^+$	432>199 [ethoxynaphtyl-carbonyl]$^+$
	Amoxycillin	366>208 $[M\text{-}NH_3+H]^+$	366>113 $[F1+H]^+$
	Oxacillin	419>144 [phenylisoxazolyl+H]$^+$	419>243 [aminodimethyl-pyridine+H]$^+$
	Penicillin G	352>160 $[F1+H]^+$	352>176 $[F2+H]^+$
	Penicillin V	368>114 $[F1\text{-}CO_2+H]^+$	368>160 $[F1+H]^+$

Table 3 (continued)

Group of substances	Compound	MRM 1	MRM 2
Other antibiotics	Chloramphenicol	323>152 [nitrobenzyl alcohol carbanion]$^-$	323>176 [194-H_2O]$^-$
	Trimethoprim	291>230 [M-2CH_3O]$^+$	291>213 [M-trimethoxy-phenyl]$^+$
β-blockers	Atenolol	267>190 [M-H_2O-NH_3-isopropyl+2H]$^+$	267>145 [190-CO-NH_3]$^+$
	Sotalol	273>255 [M-H_2O+H]$^+$	273>213 [M-C_3H_9N+H]$^+$
	Metoprolol	268>133 [$C_6H_{15}NO_2$]$^+$	268>159 [$C_8H_{17}NO_2$]$^+$
	Propranolol	260>116 [(*N*-isopropyl-*N*-2-hydroxypropyl-amine)+H]$^+$	260>183
Other drugs	Salbutamol	240>166 [M+H-(CH_3)2C-CH_2-H_2O]$^+$	240>148 [166-H_2O]$^+$
	Ranitidine	315>176 [M-$C_8H_{12}NO$]$^+$	315>130 [M-$C_8H_{12}NO$-NO_2]$^+$
	Omeprazole	346>136 [M-H_3CO-($C_7H_4N_2$)-SO-CH_2]$^+$	346>198 [M-H_3CO-$C_7H_4N_2$]$^+$

are available using Q-TOF analysers [152], but this technique has not been routinely employed yet.

In most cases, the base peak selected for quantitation of estrogens in SIM and MRM modes, when operating with an ESI (NI) and APCI (PI) interface, corresponds to the deprotonated molecule $[M-H]^-$ and to the $[M+H-H_2O]^+$ ion ($[M+H]^+$ for estrone). In Table 3, the most common fragmentations monitored in LC-MS/MS analysis, using triple quadrupole instruments, are summarized for the most studied steroid estrogens.

4.2.2
Pharmaceuticals

A large number of reports and reviews are devoted to the occurrence, fate and risk assessment of pharmaceuticals in the environment [92, 93, 127, 189–

193]. While their occurrence in the aquatic environment has been extensively studied, data regarding their presence in solid samples are still scarce, veterinary antibiotics being the ones most commonly investigated in such matrices [194–199].

Most of the analytical methods available in the literature are focused on the analysis of particular therapeutic groups. However, the general trend in recent years is the development and application of generic methods that permit simultaneous analysis of multiple-class compounds [2, 99, 200–209]. Multi-residue methods provide wider knowledge about their occurrence, necessary for further understanding of their removal, partition and ultimate fate in the environment. Nevertheless, simultaneous analysis of compounds from diverse groups with different physico-chemical properties requires a compromise in the selection of experimental conditions for all analytes studied.

4.2.2.1
Sample Preparation

In such multi-residue methods, simultaneous extraction of all target analytes in one single SPE step from water samples is the approach most widely employed [190]. Another option consists of the combination of two SPE materials operating either in series or classifying target compounds into two or more groups, according to their physico-chemical properties [190]. In both situations Oasis HLB or C_{18} cartridges are the most widely employed materials for pre-concentration and extraction of target compounds. For the former, neutral sample pH is advisable to achieve good recoveries for all compounds, whereas for C_{18}, sample pH adjustment prior to extraction is required depending on the acidic, neutral or basic nature of the analytes. The less common cartridges employed are Lichrolut ENV+, Oasis MCX and StrataX. While these materials generally need sample pH adjustment and sometimes special elution conditions (mixtures of methanol/ammonia, acidified or basified methanol), Oasis HLB provides good performances at neutral sample pH and elution with pure organic solvents, generally methanol (see Table 2).

When these methods include the determination of antibiotics, some precautions have to be taken into account during the analytical procedure. As tetracycline, sulphonamides and polypeptide antibiotics form complexes with metal ions, the addition of some chelating agent before SPE, such as Na_2EDTA, is recommended to avoid important losses during analysis. When analysing tetracycline, it should be highly recommended to use PTFE instead of glass materials, since they tend to bind to the glass, resulting in significant losses [93, 189, 190]. Additional problems are the formation of keto–enol tautomers in alkaline aqueous solutions [210] and the formation of 4-epimer isomers in acidic ones [211]. For this reason, it is advisable to work at neutral sample pH.

MIPs and immunosorbents could be a useful tool to provide high selectivity for target analytes when performing single group analysis. Although these materials have been widely employed to selectively isolate clenbuterol, aniline β-agonists, tetracycline and sulphonamide antibiotics, β-agonists and β-antagonists from biological samples, few applications have been reported for environmental matrices [212–215].

With regard to their analysis in solid samples, most of the methods available in the literature are based on sonication and PLE as the extraction technique followed by a clean-up procedure. The extraction solvents used generally consist of pure organic solvents, such as methanol and acetonitrile, or mixtures of polar solvents with water, acidified water (acetic acid, orthophosphoric acid), or buffers (citric acid) in different proportions. An important issue to consider is that when extracting tetracycline and macrolide antibiotics by PLE, temperature control is required, since temperatures higher than room temperature can cause their transformation into epi- or anhydrous forms for TCs. Moreover, values higher than 100 °C promote the degradation of macrolides [127].

For the extraction of tetracycline antibiotics, special precautions have to be taken into account. As they tend to form complexes with metal ions, extraction solvents consist of mixtures with organic solvent, generally methanol, with citric acid and McIlvaine buffer (mixture of citric acid with Na_2HPO_2), also containing Na_2EDTA [194].

After extraction, a purification step is required and is generally performed by SPE, using the same cartridges and conditions as the analysis of pharmaceuticals in water samples. Sample extracts are therefore diluted with an appropriate volume of MilliQ water, until the organic solvent content is below 10%, in order to avoid losses of target compounds during SPE [194]. Cartridges mainly used consist of Oasis HLB (see Table 2). However, some authors use either SAX or MCX [189] cartridges in tandem with the polymeric Oasis HLB [194], in order to remove negatively charged humic material (in the SAX material) and organic matter (in the MCX cartridge), and therefore selectively retain target compounds in the Oasis HLB material. When SAX cartridges are employed, samples are acidified at pH values ranging from 2 to 3 to ensure an efficient removal of the humic material (see Table 2).

Elution of target compounds from SPE cartridges is achieved with a large variety of organic solvents, according to the physico-chemical properties of the compounds analysed, methanol and acetonitrile being the most common ones (see Tables 1 and 2).

4.2.2.2
Instrumental Analysis

LC-MS/MS is the instrumental method of choice due to its versatility, specificity and selectivity, replacing GC-MS and LC-MS [190]. GC-MS can only be successfully applied for a limited number of non-polar and volatile pharmaceutical compounds, requiring a time-consuming derivatization step for the determination of polar pharmaceuticals [216–219]. Among LC-MS/MS techniques, triple quadrupole (QqQ) and ion trap (IT) instruments are in common use [92], the former being the most widely used, working in selected reaction monitoring (SRM) mode and typically reaching ng/L detection limits. More recent approaches in LC-MS/MS are linear ion traps (LITs), new generation triple quadrupoles, and hybrid instruments, such as quadrupole–time of flight (QqTOF) and quadrupole–linear ion trap (QqLIT) [92, 220].

The main applications of QqTOF instruments are focused on the elucidation of structures proposed for transformation products or are used as a complementary tool to confirm positive findings obtained by a QqQ screening method. Recently, Eichhorn et al. [221] reported on the structural elucidation of the metabolites of the antimicrobial trimethoprim. Stolker et al. [203], Marchese et al. [222], Petrovic et al. [93] and Gómez et al. [223] used QqTOF to identify the presence of various pharmaceuticals in environmental waters. Recently, Pozo et al. [224] evaluated the potential of a QqTOF instrument to confirm positive findings in the analysis of penicillin and quinolone antibiotics in surface and ground water samples. An example of the analysis of selected pharmaceuticals in an urban wastewater by UPLC-QqTOF-MS is shown in Fig. 4.

As concerns QqLIT, Seitz et al. [225] developed a method for the determination of diclofenac, carbamazepine and iodinated X-ray contrast media using direct analysis (among other contaminants), reaching LODs of 10 ng/L. Nikolai et al. [226] used QqLIT operating in QqQ mode for stereoisomer quantification of β-blockers in wastewater. On the other hand, Gros et al. [212] developed an analytical methodology for trace analysis of eight β-blockers in wastewaters using MIPs for pre-concentration of target compounds combining different functions of QqQ. Quantitative analysis was performed using a 4000QTRAP tandem mass spectrometer in SRM mode. Using the information-dependent acquisition (IDA) function in the software, a large amount of data for unequivocal identification and confirmation of the target compounds were generated at high sensitivity. An example of an IDA experiment for the determination of atenolol in an influent wastewater sample is shown in Fig. 5.

Regarding LC, reversed-phase LC is mainly used, C_{18} columns being the preferred ones. Only one method, targeted to acidic drugs, was based on ion-pair reversed-phase LC with a Phenyl–Hexyl column [227]. As mobile phases, acetonitrile, methanol, or mixtures of both solvents are normally used. In order to improve the sensitivity of MS detection and give an efficient retention, mobile phase modifiers, buffers and acids are widely employed, with ammonium acetate, tri-*n*-butylamine (TrBA), formic acid and acetic acid being the more common ones. Typical concentrations of salts range from 2 to

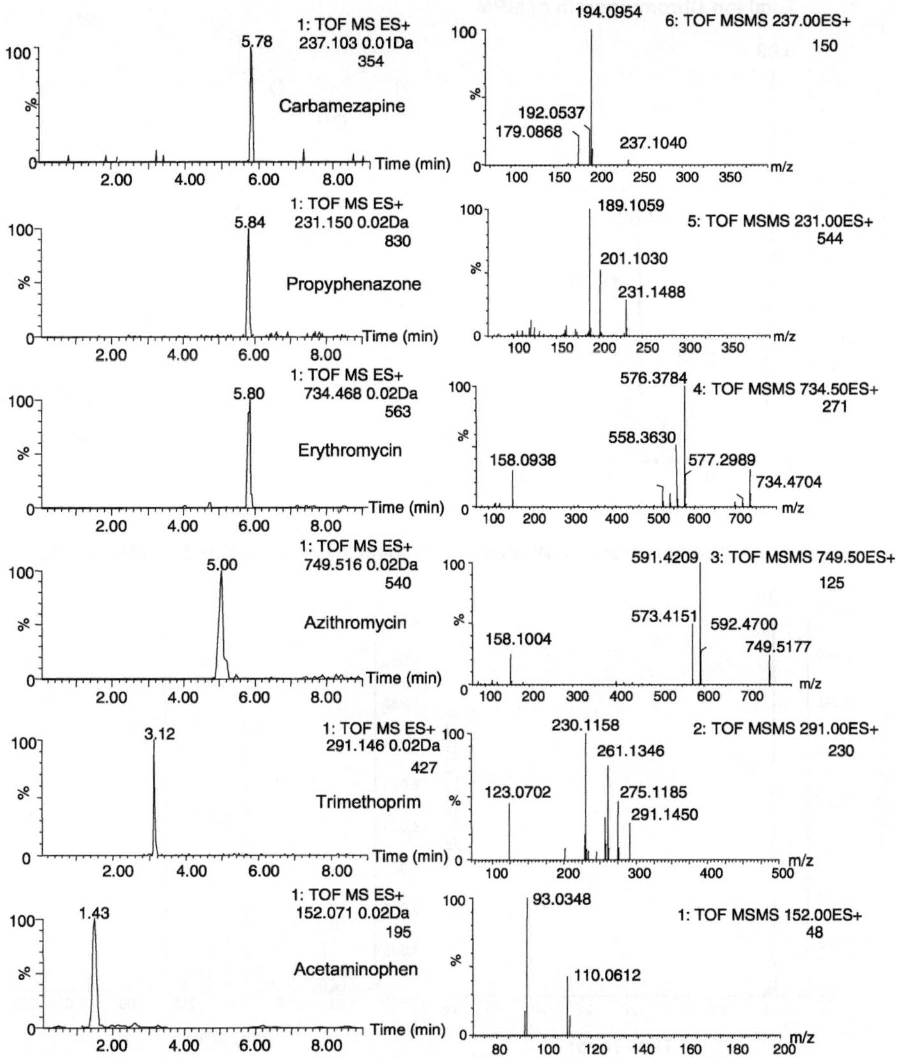

Fig. 4 Confirmation of several pharmaceuticals in an urban wastewater. *Left panel*: narrow window extracted ion chromatograms (nwXICs) of $[M+H]^+$ obtained in the TOF mode for m/z 152.071 (acetaminophen), m/z 291.146 (trimethoprim), m/z 749.516 (azithromycin), m/z 734.468 (erythromycin), m/z 231.150 (propyphenazone) and m/z 237.103 (carbamazepine). *Right panel*: product ion spectra obtained in the Q-TOF mode

20 mM, since it has been observed that higher concentrations could lead to a reduction of signal intensities [190].

Shortening the analysis time is important for attaining the high sample throughput often required in monitoring studies. This can be achieved by

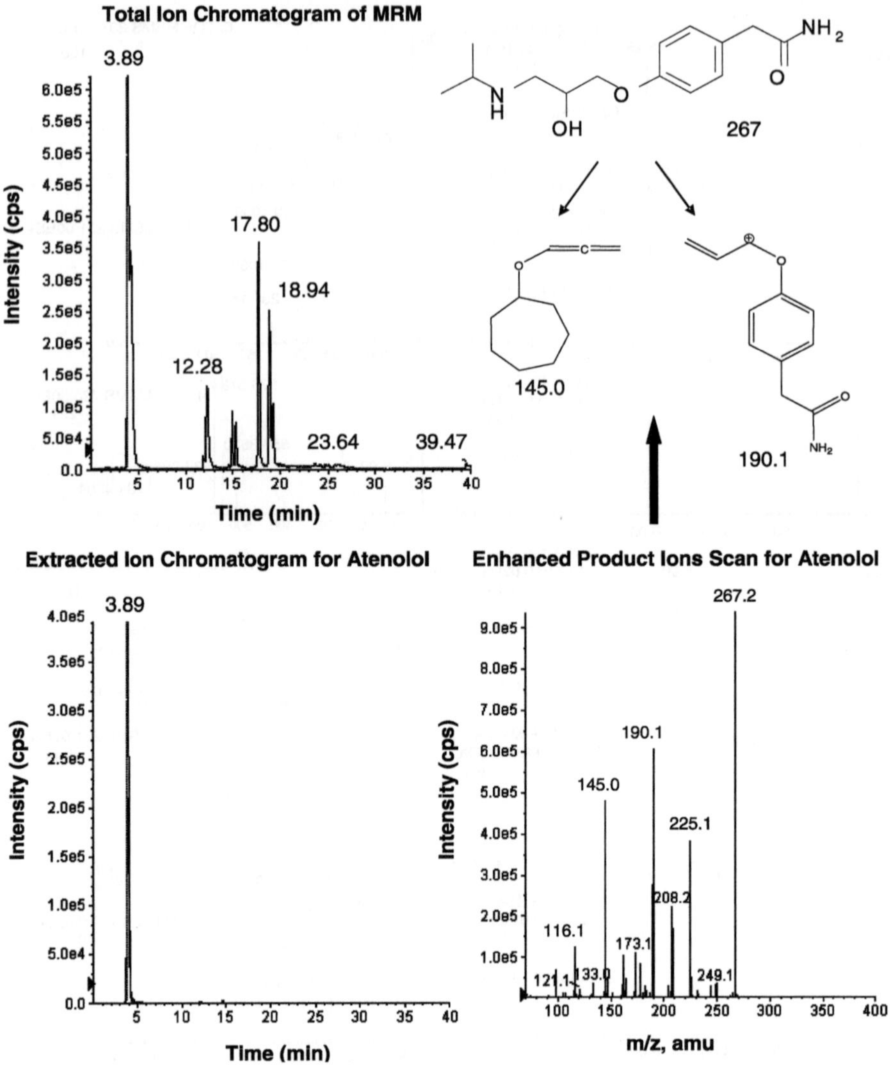

Fig. 5 Information-dependent acquisition (IDA) experiment for the determination of atenolol in an influent wastewater sample

using short columns and increased flow velocity, decreasing the particle size of stationary phases or increasing temperature. These approaches are applied in two newly developed instruments, UPLC (ultra-performance LC) and by RRLC (rapid resolution LC). For the moment, only one publication presented by Petrovic et al. [93] describes the use of UPLC coupled to a QqTOF system for the multi-residue analysis of 29 pharmaceuticals in environmental waters. Compounds more frequently detected in multi-residue methods and their MRM transitions are summarized in Table 3.

4.2.3 Personal Care Products (PCPs)

This group of compounds includes synthetic musk fragrances (nitro and polycyclic musk fragrances), antimicrobials (triclosan and its metabolites and triclocarban), sunscreen agents (ultraviolet filters), insect repellents (*N*,*N*-diethyl-*m*-toluamide, known as DEET) and parabens (*p*-hydroxybenzoic esters), which are basically substances used in soaps, shampoos, deodorants, lotions, toothpaste and other PCPs. The nitro musk fragrances were the first to be produced and include musk xylene, ketone, ambrette, moskene and tibetene. In the environment, the nitro substituents can be reduced to form amino metabolites of these compounds. The polycyclic musk fragrances, which are used in higher quantities than nitro musks, include 1,2,4,6,7,8-hexahydro-4,6,6,7,8,8-hexamethylcyclopenta-γ-2-benzopyrane (HHCB), 7-acetyl-1,1,3,4,4,6-hexamethyl-1,2,3,4-tetrahydronaphthalene (AHTN), 4-acetyl-1,1-dimethyl-6-*tert*-butylindane (ADBI), 6-acetyl-1,1,2,3,3,5-hexamethylindane (AHMI), 5-acetyl-1,1,2,6-tetramethyl-3-isopropylindane (ATII) and 6,7-dihydro-1,1,2,3,3-pentamethyl-4-(5*H*)-indanone (DPMI). Parabens are the most common preservatives used in personal care products and in pharmaceuticals and food products. This group of substances includes methylparaben, propylparaben, ethylparaben, butylparaben and benzylparaben.

These substances have been analysed in various environmental matrices, such as water, sediments, sewage sludge and aquatic biota. The hydrophobicity of many of these compounds indicates their potential for bioaccumulation [228].

4.2.3.1 Sample Preparation

Methods used for the extraction of PCPs from water samples are based on liquid–liquid extraction (LLE) [1, 52–67], continuous liquid–liquid extraction (CLLE), SPE [219, 229–231] and SPME [232, 233]. When LLE and CLLE are applied, various organic solvents are used for the extraction of target compounds, dichloromethane, pentane [234, 235], hexane [236–238], toluene [239, 240], cyclohexane [233] and petroleum ether [241], and mixtures of them in appropriate proportions, being the most widely employed (see Table 2). Extraction of target compounds using these techniques is performed either at ambient pH or by acidifying the sample, generally to values ranging from pH 2 to 3 [219, 228]. For the extraction of UV filters, LLE with cyclohexane at pH 3 is the most common procedure [228].

For SPE, a wide range of sorbents are used, including C_{18} [219, 230, 231, 242–248] at ambient and acidic (pH<3) sample pH, Abselut Nexus [249, 250] (Varian, Palo Alto, CA, USA), Isolute ENV+ [231], Oasis MAX [241], Bio Beads

SM-2 [251–253] (Bio-Rad Laboratories, Hercules, CA, USA), XAD-2 [254] (Supelco, St. Louis, MO, USA), SDB-XC [255, 256] and XAD-4/XAD-8 [254, 257]. Elution of target compounds from these materials is achieved with a large variety of organic solvents, according to the physico-chemical properties of the compounds analysed, with acetone, methanol, toluene, hexane, mixtures of dichloromethane/acetone and methanol, hexane/acetone or hexane/ethyl acetate and acetone/ethyl acetate being the most widely used [228]. When analysing antimicrobials with Oasis MAX, the sample is acidified (pH 3) prior to extraction, washed with methanol/sodium acetate solution and eluted with pure methanol. For parabens, few methods are reported relevant to environmental matrices, but their analysis is mainly based on SPE extraction using Oasis HLB.

Sometimes, when using these techniques, sample purification prior to instrumental analysis is necessary, generally using SPE with silica and alumina [228]. The most common techniques used for their extraction from sewage sludge include PFE [197, 231, 241, 244, 245, 252, 258, 259], SFE [230, 241] (using CO_2), sonication, Soxhlet [240, 260–263], LLE [264, 265] and MAE [266]. Sometimes, before extraction of target compounds, copper is added to remove sulphur content in the samples. Generally, after extraction, a purification step with silica columns or size-exclusion chromatography (SEC) followed by Bio Beads SX-3 or silica columns is required. Hexane, ethyl acetate, acetone, cyclohexane and mixtures of them are the solvents mainly used for the elution of target compounds [228].

On the other hand, SPME has also been a widespread technique for the extraction of PCPs in environmental waters and solid samples, using either direct (DI-SPME) or headspace (HS-SPME) methods [228, 248, 267, 268]. The materials most commonly used are polydimethylsiloxane (100 μm) (PDMS) for DI-SPME, and PDMS-DVB (65 μm), Carboxen-PDMS (75 μm), Carbowax-DVB (65 μm) and Carbowax-PDMS (65 μm) for both types of SPME, PDMS-DVB being the one yielding higher recoveries [228].

The extraction techniques used for the analysis of biota samples are the same as those used for solid samples but after extraction, removal of the lipid content is essential, generally performed by SEC in tandem with Bio Beads SX-3 cartridges. For the determination of nitro musks, lipids cannot be removed destructively with H_2SO_4 since important losses of target compounds could occur.

4.2.3.2
Instrumental Analysis

Synthetic musk fragrance standards and deuterated musk xylene and AHTN standards are commercially available for use as recovery or injection standards. There have been reports of problems with the use of the deuterated AHTN (AHTN-d_3) due to the occurrence of proton exchange during sample processing [228]. A variety of other recovery and injection standards have been used for the analysis of synthetic musk fragrances, including pentachloronitrobenzene, deuterated polycyclic aromatic hydrocarbons (PAHs), and various labelled and unlabelled polychlorinated biphenyls (PCBs).

PCPs are most commonly analysed by GC-EI-MS, but GC-NCI-MS is more sensitive for nitro musk fragrances. These compounds have also been analysed by GC-FID, GC-ECD, and high-resolution and ion-trap tandem mass spectrometry (MS/MS). Common GC phases are 5% phenylmethylpolysiloxane and dimethylpolysiloxane [228].

Triclosan and its chlorinated metabolites are also determined by GC-EI-MS with and without derivatization, LC-MS and LC-MS/MS. When derivatizing, *N*,*N*-diethyltrimethylamine (TMS-DEA), *N*,*O*-bis(trimethysilyl)trifluoroacetamide (BSTFA), pentafluorinated triclosan and *tert*-butyldimethylsilyl triclosan are the ether derivatives generated after reaction with methyl chloroformate (MCF), pentafluoropropionic acid anhydride (PFA) and *N*-*tert*-butyldimethylsilyl-*N*-methyltrifluoroacetamide (MTBSTFA), respectively [228].

GC-based techniques dominate the analysis of UV filters and insect repellents, using DB-5 and 5% polyphenylmethylsilicone columns, respectively. Almost all UV filters are amenable to GC except octyl triazone, avobenzone, 4-isopropyldibenzoylmethane and 2-phenylbenzimidazole-5-sulphonic acid, some of them being determined by HPLC-UV. Although there are few methods published dealing with the analysis of parabens in environmental samples, the methods reported are based on LC-MS/MS under NI conditions using a C_{18} column.

4.3 Surfactants

A number of books and reviews are already available on the determination of surfactants in wastewaters, sludges, sediments and biological samples, using GC-MS, LC-MS or LC-MS/MS techniques [4, 269–271]. Among the various surfactant classes, both non-ionic and ionic substances are the most widely employed in both industry (e.g. alcohol ethoxylates (AEOs), alkylphenol ethoxylates (APEOs) and different fatty amine or acid ethoxylates [269]) and household applications (linear alkylbenzene sulphonates (LASs)).

From the environmental point of view, APEOs and LASs are the ones deserving especial attention due to their ubiquity and ecotoxicological relevance. Sixty percent of APEOs that enter mechanical or biological sewage or sewage sludge treatment plants are subsequently released into the environment, 85% being in the form of the potentially estrogenic metabolic products, alkylphenols (APs), alkylphenol carboxylates (APECs) and alkylphenol dicarboxylates (CAPECs) [272–275]. Moreover, numerous studies have confirmed that alkylphenolic compounds can mimic endogenous hormones. APEOs and their biodegradation products are transformed into halogenated by-products during chlorination disinfection in wastewater or drinking water treatment plants, in the presence of bromide ion [276, 277].

4.3.1 Sample Preparation

Both ionic and non-ionic surfactants are generally isolated from water samples by SPE, at natural sample pH, Lichrolut C_{18} cartridges (Merck, Darmstadt, Germany) being the most widely employed. For halogenated derivatives, SPE using Lichrolut C_{18} is also widely employed [278]. Elution is usually performed using pure solvents, with methanol the most common one [5].

Analysis of surfactants and their halogenated derivatives from solid samples is challenging due to their strong adsorption on the soil/sludge particles by hydrophobic and electrostatic interactions. Most of the methods available in the literature are based on sonication and PLE as the extraction technique followed by a clean-up procedure, generally using SPE C_{18}, ENV+, strong anion exchange (SAX) or polymeric cartridges [5, 279–281]. The former has been widely employed for the analysis of LASs, NPEOs and their degradation products nonylphenol carboxylates (NPECs) and NPs, AEOs, and coconut diethanolamides (CDEAs) [282]. On the other hand, PLE methods have been optimized for LASs, NPEOs and their neutral and acidic metabolites, AEOs and alkylamine ethoxylates (ANEOs) [282]. Pure solvents, such as methanol and dichloromethane, and mixtures of organic solvents (hexane/acetone or methanol/dichloromethane) are mainly used for the extraction of surfactants from solid matrices (see Table 2). Other methods based on extraction with pressurized (supercritical) hot water as well as SFE with solid-phase trapping, using CO_2 and methanol or water as modifier, have been described in the literature for the simultaneous extraction of several surfactant classes [282].

4.3.2 Instrumental Analysis

Commercial mixtures of surfactants comprise several tens to hundreds of homologues, oligomers and isomers. For LASs, mixtures of secondary isomers with alkyl chain lengths of 10–13 carbons are available.

GC and LC coupled to MS detection systems are now the commonly used methods to identify and quantitate surfactants in both aqueous and solid matrices. Although GC-MS is adopted in many analytical methodologies, it cannot be applied for the direct determination of several classes of surfactants since derivatization of low volatility compounds is required. This is why, in surfactants analysis, GC-MS methods are partially substituted with LC-MS or LC-MS/MS [269, 283]. However, most of the methods available focus on one or two classes of surfactants which are similar in nature, generally including their main degradation products. Only recently, several efforts have been made to develop generic methods that allow simultaneous determination of a broad range of surfactant types.

Gas chromatography–mass spectrometry has been widely used for the analysis of alkylphenolic compounds and anionic surfactants (LASs). Alkylphenolic substances, which mainly include the most volatile compounds AP, APEO, AEO and ANEO with fewer than four ethoxy groups, and the rest of the non-ionic surfactants can be determined without derivatization, while for anionic surfactants derivatization prior to analysis is required [284]. Derivatization is usually performed by transforming parent compounds to the corresponding trimethylsilyl ethers, methyl ethers, acetyl esters and pentafluorobenzoyl or heptafluorobutyl esters [5, 285, 286]. After derivatization, NPEO derivatives can be analysed by GC-MS in the EI or NCI modes [130]. GC-CI-MS, using ammonia as reagent gas for the detection of NPE_nC, gave intense ammonia–molecular ion adducts of the methyl esters, at *m*/*z* 246, 310, 354 and 398 for NPE_1C, NPE_2C, NPE_3C and NPE_4C, respectively, with little or no secondary fragmentation [5]. Moreover, GC-CI-MS spectra of the NPECs with isobutene as reagent gas showed characteristic hydride-ion-abstracted fragment ions shifted 1 Da from those in the corresponding EI mass spectra. On-line direct GC injection-port derivatization with ion-pair reagents (tetraalkylammonium salts) has also been reported [287].

As concerns liquid chromatography, even though LC-MS/MS is more specific and sensitive than LC-MS, the majority of studies dealing with the analysis of surfactants in environmental samples are based on LC-MS [128, 270]. However, several papers describing the application of tandem MS to the unambiguous identification and structural elucidation of alkylphenolic compounds have been published [275, 288–291].

The analysis of LASs by LC-MS operating in the ESI and NI modes is particularly attractive due to their anionic character. MS analysis of commercial LAS mixtures shows four ions at *m*/*z* 297, 311, 325 and 339, corresponding to deprotonated molecules of C_{10}–C_{13} LAS homologues [282]. With increasing cone voltage using in-source collision-induced dissociation (CID), the spectra show additional fragment ions at *m*/*z* 183 and 80, which were assigned to styrene-4-sulphonate and $[SO_3]^-$. The analysis of APEOs by LC-MS in the PI mode yields a characteristic pattern of equally spaced signals with mass differences of 44 Da (one ethoxy unit), which is a diagnostic fingerprint for this group of compounds. Using an ESI interface and aprotic solvent, APEOs predominantly give evenly spaced sodium adducts $[M + Na]^+$ [270], which are relatively stable and generally no further structurally significant fragmentation is provided in the mass spectrum. Some authors used ammonium acetate as mobile phase in order to enhance the formation of ammonium adducts over sodium or proton adducts, which give fragments in CID processes, enabling a more specific detection of APEOs [275].

On the other hand, alkylphenoxy carboxylates (APE_nC) are generally determined by ESI operating in the NI mode, and less frequently by the PI mode [282]. For the analysis by NI, two types of ions, one corresponding to the deprotonated molecule and the other corresponding to deprotonated alkylphenols, are obtained. For the determination of AEOs, some authors used LC-MS operating in APCI mode [282] to analyse AEOs with alkyl chains from C_{10} to C_{14} and from C_{10} to C_{18}.

Like their non-halogenated analogues, halogenated APEOs show a great affinity for alkali metal ions when analysed by LC-MS in ESI mode, and they give exclusively evenly spaced (44 Da) sodium adduct peaks $[M+Na]^+$ with no further structurally significant fragmentation [277]. Fully de-ethoxylated degradation products, octylphenol (OP) and nonylphenol (NP), were detected under NI conditions with both APCI and ESI interfaces. However, sensitivity was higher when using an ESI source than an APCI one [5].

Diagnostic ions used for the analysis of XAPEOs under NI conditions using LC-MS corresponded to the cleavage of the alkyl moiety (CH_2 group), leading to a sequential loss of *m*/*z* 14, the most abundant fragments being at *m*/*z* 167 for ^{35}Cl and *m*/*z* 169 for ^{37}Cl.

In LC-tandem MS, compounds analysed under NI conditions (AP, APEC and their halogenated derivatives) were analysed by ESI-MS/MS, while for APEO, detected in the PI mode, no fragmentation was obtained using an ESI source. These compounds were determined by APCI-MS/MS. Using ESI-MS/MS, the CID spectrum of NP shows fragments at *m*/*z* 147, 133, 110 and 93, attributed to the progressive fragmentation of the alkyl chain [5]. For APEC, an intense signal at *m*/*z* 219 is observed for NPEC, produced after the loss of the carboxylated (ethoxy) chain, and other peaks at *m*/*z* 133 and 147, due to the sequential fragmentation of the alkyl chain [128, 275, 288]. LC-tandem MS was also used to determine halogenated surfactants, obtaining the same product ions as for LC-MS, with *m*/*z* 167 for ^{35}Cl and *m*/*z* 169 for ^{37}Cl, with a relative ratio of intensities of 3.03, being the most abundant fragment ions.

LC-ESI-IT-MS and LC-(PI)-APCI-IT-MS have been used to determine LASs and SPCs, and APEOs, AEOs and cationic surfactants, respectively, in several environmental matrices [292–296]. These instruments permit MS^n, which makes them suitable for identification and quantitation purposes. On the other hand, MALDI-TOF and MALDI-Q-IT have been used to determine APEOs [297, 298]. Ayorinde et al. [292] used α-cyano-4-hydroxycinnamic acid as a matrix to determine NPEO (with 2–120 ethoxy units).

4.4 Polybrominated Diphenyl Ethers (PBDEs)

Polybrominated flame retardants are chemicals used in large quantities as they are added to polymers, which are used in plastics, textiles, electronic circuitry and other materials, to prevent fires, due to their fire retarding properties [299]. Several studies have reported that these substances tend to bioaccumulate in biota and humans due to their lipophilicity [300–311]. Moreover, PBDEs are suspected to cause endocrine dysfunction by interfering with thyroid hormone metabolism [312, 313]. In 2003, the European Union banned the use of the PBDE commercial mixtures PentaBDE and OctaBDE. Nowadays, the only remaining unregulated PBDE mixture in production is DecaBDE [314].

4.4.1
Sample Preparation

Analytical methods developed for the determination of PBDEs are very similar to those used for PCBs, due to their similarity in physico-chemical properties. As they are non-polar compounds, their occurrence has been widely reported in solid samples, such as sewage sludge, soil and sediments. For this reason, the determination of PBDEs in liquid samples is mainly focused on the analysis of human milk or plasma, while few studies have analysed them in natural and sewage waters [81].

BDE congeners typically measured in human tissues are associated primarily with the PentaBDE mixture, and to some extent with the OctaBDE mixture. One of the greatest challenges to measuring PBDEs in environmental samples has been developing methods to accurately quantify BDE 209. While analytical methods are readily available for quantifying tribrominated through heptabrominated congeners found in the PentaBDE and OctaBDE mixtures, the analysis of brominated compounds has proven to be difficult. Currently, there are several reviews available in the scientific literature devoted to the analysis of PBDEs in different environmental matrices [81, 82, 299].

The techniques used are mainly based on liquid–liquid extraction (LLE) [315–319], with mixtures of non-polar and polar solvents. Recently, headspace solid-phase microextraction (HS-SPME) and microporous membrane liquid–liquid extraction (MMLLE) have been proposed as suitable techniques [320]. Other techniques used consist of saponification with ethanolic KOH, especially for their analysis in human milk [299]. Similar procedures involving protein denaturation with HCl/isopropanol and extraction with hexane/methyl *tert*-butyl ether have been used for the determination of neutral and phenolic brominated compounds from human serum [321].

Extraction of PBDEs from solid and biological samples is generally performed using non-polar solvents, such as hexane, toluene, dichloromethane or hexane/acetone mixtures. Binary solvent mixtures, combining a non-polar and a polar solvent, are most commonly used for their known extraction efficiency, especially for biota and wet sediment samples, as non-polar solvents are not able to penetrate the organic matter and therefore desorb contaminants. Soxhlet [322–324], supercritical-fluid extraction (SFE) [325], accelerated solvent extraction [326, 327] and microwave-assisted extraction (MAE) are the techniques mainly used [328].

Extracts obtained using these techniques need a clean-up step prior their analysis by chromatographic techniques. Therefore, extracts from sediments, sewage sludge or soil samples may contain sulphur that has to be removed as it could disturb the GC analysis. Typical methods used for this purpose are treatment with copper powder, silica modified with $AgNO_3$ in a multilayer silica column, desulphuration with mercury or reaction with tetrabutylammonium sulphite [81, 82, 299]. In the case of Cu powder, it is generally added in the Soxhlet beaker or PLE cell.

On the other hand, in the case of sewage sludge, extracts contain a high amount of lipids and organic matter, which should be removed prior to instrumental analysis, by either non-destructive or destructive methods. The former include gel permeation and column adsorption chromatography, using polystyrene–divinylbenzene copolymeric columns and dichloromethane or mixtures of dichloromethane/hexane and ethyl acetate/cyclohexane as eluents. Other neutral adsorbents commonly used are silica gel, alumina and Florisil® [323, 329]. Destructive lipid removal methods consist of sulphuric acid treatment, either directly to the extract or via impregnated silica columns, and saponification of extracts by heating with ethanolic KOH. Since PBDE concentrations are generally related to the amount of lipids, the lipid content is often measured gravimetrically prior to the clean-up step, or determined separately by a total lipid determination [299, 323].

It is important to remark that when analysing BDE 209 special precautions should be taken, as it is sensitive to UV light and it may also adsorb to small dust particles. Therefore, incoming sunlight into the laboratory should be blocked and all glassware covered with aluminium foil, to prevent dust particles and UV light entering either the solutions or samples. The use of isooctane for the extraction should be avoided due to the insolubility of BDE 209 in this solvent. Moreover, it is recommended not to evaporate extracts until dryness because it may not completely re-dissolve after that step even when using toluene.

4.4.2
Instrumental Analysis

Like perfluorinated alkyl substances, standards available for PBDE determination consist of a mixture of several congeners of different degrees of bromination. As reported by Stapleton [314], about 160 of the 209 possible BDE congeners are currently commercially available. Isotopically labelled standards to be used for internal standard calibration purposes are scarce, and therefore some authors have used ^{13}C-labelled bromobiphenyls and chlorinated diphenyl ethers as an alternative.

Owing to their vapour pressure and polarity, GC coupled to ECD, NCI-LRMS and EI-LRMS detectors has become a standard analytical separation method for the analysis of PBDEs. The three most common injection techniques for PBDEs are split/splitless, on-column and programmable temperature vaporization (PTV) injection. When working with split/splitless injection, the high inlet temperature can lead to thermal degradation and discrimination of higher molecular weight PBDEs, particularly the fully brominated BDE 209. This problem can be solved by using on-column injection, which consists of the direct injection of the sample, dissolved in a carrier solvent, onto the head of the column [314, 330]. PTV inlets have become a more popular choice for injection over the past 5 years, where higher injection volumes can be used, thus improving detection limits.

Both on-column and PTV injections require the use of a guard column, composed either of untreated silica with active silanol groups or deactivated fused silica. Short DB columns (10–15 m) with thin (0.1 μm) stationary phases are the most commonly used and the ones providing higher sensitivity for measuring the entire range (low to high bromine substitution). However, longer columns are not well suited for higher molecular weight PBDEs, as they can degrade [314]. Again, BDE 209 should receive special attention, due to its susceptibility to degrade at higher temperatures in the GC system.

ECNI-LRMS provides higher sensitivity than EI-LRMS, the LODs for the former being at least one order of magnitude lower than for the latter. However, EI-LRMS provides higher specificity and accuracy in quantification, as isotopically labelled standards can be used for the isotope dilution approach.

GC/ECNI-LRMS mass spectra for all PBDEs rely upon selective ion monitoring (SIM) of Br^- ions [^{79}Br and ^{81}Br]. By contrast, EI provides more structural information, giving the molecular ions and the sequential losses of bromine atoms (molecular clusters for mono- to tri-BDEs and $[MBr_2]^+$ for tetra- to hepta-BDEs).

The presence of potential interferences in the NCI and EI approaches has been widely studied [314, 331, 332]. In general, EI-MS is affected by chlorinated interferences, especially PCBs, as analytical procedures developed for PBDE analysis are mainly based on the methods already available for PCBs. Thus, purified extracts may contain both PCBs and PBDEs. Alaee et al. [332] found that the isotopic cluster of $[M-Cl_2]^+$ from heptachlorinated biphenyls contains the same mass fragments found in tetrabrominated diphenyl ethers $[M-Br_2]^+$ and resolving powers of 25 000 ($m/\Delta m$) were required to differentiate them.

Such interferences are illustrated in Figs. 6 and 7, where the chromatograms obtained following the injection of a PBDE standard mixture and PCB standard mixtures are depicted. As can be observed, some hepta-CBs (CB-180) and octa-CBs (CB-199) elute with tetra-BDEs. Furthermore, some octa-CBs (CB-194) elute with penta-BDEs [82].

When using NICI-LRMS, such chlorinated interferences do not occur, but due to the presence of different brominated compounds, such as MeO-BDEs,

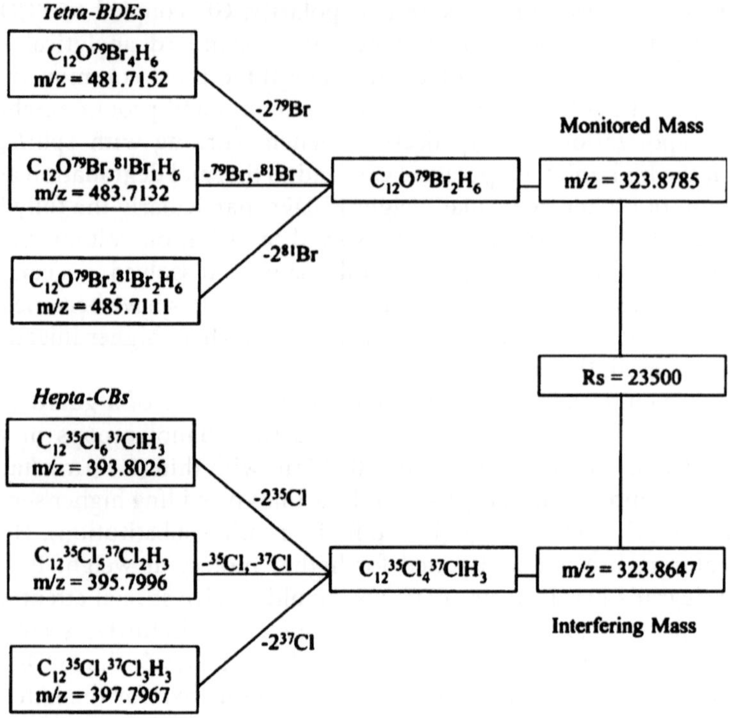

Fig. 6 Interferences between tetra-BDEs and hepta-CBs. Reprinted with permission from Elsevier [331]

can produce the same fragment ion and confound analysis of PBDEs. Several papers have reported the co-elution of 2,2′4,4′,5′5-hexabromobiphenyl (PBB 153) and TBBPA with BDE 154 and of tetrabromobisphenol A with BDE 153 [81, 323, 333–336] on 15- and 30-cm capillary columns. Moreover, naturally produced brominated compounds, such as halogenated bipyrroles and brominated phenoxyanisoles, can be considered as potential interferences.

High-resolution instruments operating in the EI mode offer the best selectivity for PBDE measurements, with a mass resolution of approximately 10 000, resulting in fewer co-eluting interferences [337]. Moreover, they also allow the use of isotope dilution with ^{13}C-labelled BDE standards due to the reduction of interferences.

Tandem mass spectrometers using ion traps have also been reported for the analysis of PBDEs [338, 339], offering the advantage of increased sensitivity at low mass resolution because analytes are fragmented twice, minimizing the chance of isobaric interferences and reducing background noise. In this equipment, precursor ions, which are typically $[M]^+$ or $[M-Br_2]^+$, are fragmented yielding $[M-COBr]^-$ ions.

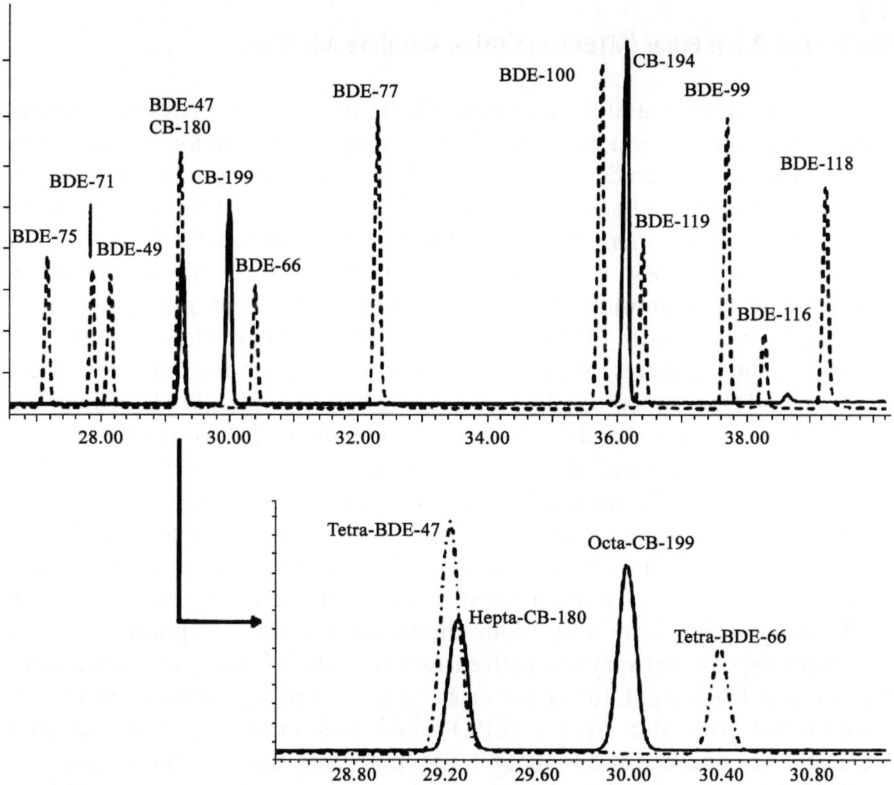

Fig. 7 TIC obtained following the co-injection of PBDE and PCB standard mixtures. Hepta- and octa-CBs eluted within the chromatographic window are defined for tetra- and penta-BDEs. BDE-47 and CB-180 eluted at the same retention time. Reprinted with permission from Elsevier [331]

HR-TOF mass spectrometers have also been used to determine PBDEs in environmental samples, with detection limits comparable to those of most other MS techniques [340, 341]. Alternative analytical techniques are LC-MS, LC-MS/MS [342, 343] and GC×GC [336, 340]. The former two are promising, but use atmospheric pressure photoionization (APPI), as PBDEs do not ionize well with either ESI or APCI. When working with APPI, both negative and positive ionization modes are suitable for their analysis, depending on the degree of bromine substitution. However, the analysis of metabolites, such as hydroxylated BDEs (OH-BDEs), can be successfully conducted when operating in ESI mode. Finally, GC×GC could be very useful to avoid the co-elution problems found in standard GC-MS methods [344].

4.5 Methyl *tert*-Butyl Ether (MTBE) and Other Gasoline Additives

MTBE, and gasoline additives in general, are not usually analysed in wastewaters, but this section was included as they are an important group of compounds to be considered when dealing with emerging contaminants. Fuel oxygenates have been added to gasoline since the 1970s, mainly as octane enhancers that increase the combustion efficiency and reduce toxic air emissions, such as lead compounds or carbon monoxide. Since the ban on tetraalkyl lead compounds, MTBE has become the most commonly used oxygenate and the one with the highest production volume worldwide [345].

Among fuel additives, MTBE is the ether with higher solubility and lower sorption and Henry's law constant, enhancing its higher mobility (nearly as fast as that of ground water) and the difficulty in removing it from water by aeration or degradation processes [346]. For this reason, as well as its intense use, MTBE has become one of the most frequently detected volatile organic compounds (VOCs) in ground water which can be adsorbed on subsurface solids [346].

Besides the health effects, toxicity and carcinogenicity at high concentrations [347], there is much interest in the aesthetic implications of MTBE in drinking water. Taste and odour thresholds for this compound in water have been reported at very low concentrations, approximately 25–60 μg/L for flavour and 40–70 μg/L for odour at 25 °C [347]. For this reason, the US Environmental Protection Agency (EPA) established a drinking water advisory for aesthetic concerns at 20–40 μg/L [347]. To date, there are no regulations for MTBE in water, air or soil in Europe but some countries are establishing their own guidelines.

Analytical methodologies dealing with the analysis of MTBE also include the determination of its main degradation products, *tert*-butyl alcohol (TBA) and *tert*-butyl formate (TBF), as well as other gasoline additives present in fuel, such as the oxygenate dialkyl ethers, for example ethyl *tert*-butyl ether, *tert*-amyl methyl ether and diisopropyl ether, and the aromatic compounds benzene, toluene, ethylbenzene and xylene (BTEX).

4.5.1 Analysis in Environmental Samples

There are some reviews devoted to the analysis of MTBE and other gasoline additives in environmental samples [346, 348, 349]. Even though MTBE is more likely to be present in ground and surface waters as well as soil samples, due to its physico-chemical properties (high mobility and solubility), some studies also revealed its presence in wastewaters [350, 351].

The most crucial step in trace analysis of VOCs is definitely enrichment and sampling. For MTBE analysis, samples do not need to be preserved, as biodegradation is very slow [352]. However, special precautions have to be taken in VOC analysis to avoid losses and prevent contamination. Bottles used to collect samples are filled to the top, avoiding air bubbles passing through the sample, to prevent volatilization of target compounds [347].

As to enrichment techniques, some methodologies, including direct aqueous injection (DAI), membrane-introduction mass spectrometry (MIMS), headspace (HS) analysis, purge and trap (P&T), solid-phase microextraction (SPME) by direct immersion or headspace compound-specific stable isotope analysis (CSIA), which is an emerging tool in environmental sciences, have been proposed and discussed by [353, 354] as appropriate methods to be used. These techniques are recommended when VOCs are found at lower concentrations and they mainly operate coupled to an instrumental technique. As VOCs, fuel oxygenates are almost exclusively analysed by GC and MS detection. Other detectors, such as flame ionization (FID), photoionization (PID) and atomic emission (AED), can also be used, but MS is the preferred one due to its higher sensitivity and selectivity [350]. In Tables 1 and 2, some of the most representative methods for the analysis of MTBE and other gasoline additives in water and solid samples, respectively, are described.

The selection of one technique or another depends on the type of matrix analysed, the concentration range and the need for compliance with the regulations [350]. P&T and SPME were the methods that obtained the best accuracy in a MTBE inter-laboratory study with 20 European participating laboratories and, when coupled with mass spectrometry, were the ones offering the best results according to the quality state assurance/quality control requirements [350, 355]. When P&T is used, VOCs are purged from water with helium, and generally they are subsequently adsorbed onto a Tenax® silica gel–charcoal trap. After sample loading, trapped components are desorbed at high temperatures and transferred directly to the GC-MS system [347].

For the analysis of MTBE and gasoline additives in solid samples, the same techniques as for water samples (P&T, SPME, etc.) are used [350]. Pressurized-liquid extraction (PLE) has also been used for the determination of higher concentrations (mg/kg) of BTEX (Application note 324) in soils using hexane/acetone (1:1). A semi-automatic purge-and-membrane inlet mass spectrometric (PAM-MS) instrument [377] provided good sensitivity and accuracy for some BTEX compounds and MTBE. Among the ifferent types of P&T instruments assembled for the analysis of VOCs in solid matrices [356–361], closed-system P&T is directed to determine low concentrations (<200 μg/kg), as indicated in the EPA Method 5035 [350].

Quantitative analysis of MTBE, its degradation products and other gasoline additives is performed by operating the mass spectrometer in EI mode, generally at 70 eV. In order to increase sensitivity and selectivity, samples are injected in time scheduled SIM mode. Due to the rather high energy transfer in the EI ionization mode, fuel oxygenates do not yield molecular ions. Typical fragments obtained correspond to the α-cleavage $[M - CH_3]^+$ or $[M - CH_5]^+$, taken as base peaks in the mass spectra [347]. Typical columns used in the GC separation are fused-silica capillary DB-624 columns (75 m × 0.53 mm ID) with a 3-μm film thickness.

5 Conclusions

Among modern analytical techniques, GC and LC, coupled to both MS and tandem MS, are the key techniques for the determination of emerging contaminants in complex environmental samples. These techniques, combined with appropriate sample preparation procedures, allow the detection of target compounds at the low environmental levels. Furthermore, the introduction of new chromatographic techniques, such as fast LC, fast GC, and GC×GC, has improved the analysis of complex mixtures. However, current analytical methods only focus their attention on parent target compounds and rarely include metabolites and transformation products. The question is whether chemical analysis of only target compounds is sufficient to assess contaminants present in the environment. Recent developments in the mass spectrometry field, such as the introduction of Q-TOF and Q-LIT instruments, allow the simultaneous determination of both parent and transformation products. Exact mass measurements provided by Q-TOF and the ability to combine several scan functions are a powerful tool to provide a more accurate identification of target compounds in complex samples, as well as to enable structural elucidation of unknown compounds. However, general screening for unknown substances is time-consuming and expensive, and is often shattered by problems, such as lack of standards and mass spectral libraries. Therefore, effect-related analysis, focused on relevant compounds, nowadays seems to be a more appropriate way to assess and study environmental contamination problems.

Acknowledgements This work was financially supported by the European Union EMCO project (INCO-CT-2004-509188) and by the Spanish Ministerio de Ciencia y Tecnología (EVITA project CTM2004-06265-C03-01).

References

1. Petrovic M, Gonzalez S, Barcelo D (2003) TrAC-Trends Anal Chem 22:685
2. Gros M, Petrovic M, Barcelo D (2006) Talanta 70:678
3. Shang DY, Ikonomou MG, McDonald RW (1999) J Chromatogr A 849:467
4. Petrovic M, Barcelo D (2002) Chromatographia 56:535
5. Barcelo D, Petrovic M, Eljarrat E, Lopez De Alda MJ, Kampioti A (2004) Chromatography 69B(6):987
6. Namiesnik J, Zabiegaa B, Kot-Wasik A, Partyka M, Wasik A (2005) Anal Bioanal Chem 381:279

7. Kozdron-Zabiegala B, Przyjazny A, Namiesnik J (1996) Indoor Built Environ 5:212
8. Belardi RP, Pawliszyn JB (1989) Water Pollut Res J Canada 24:1
9. Kot A, Zabiegala B, Namiesnik J (2000) TrAC-Trends Anal Chem 19:446
10. Lauridsen FS (2005) Environ Pollut 136:503
11. Lacorte S, Barcelo D (1996) Anal Chem 68:2464
12. Ferrer I, Hennion MC, Barcelo D (1997) Anal Chem 69:4508
13. Ferrer I, Pichon V, Hennion MC, Barcelo D (1997) J Chromatogr A 777:91
14. Ferrer I, Barcelo D (1999) J Chromatogr A 854:197
15. Renner T, Baumgarten D, Unger KK (1997) Chromatographia 45:199
16. Aguilar C, Ferrer I, Borrull F, Marce RM, Barcelo D (1998) J Chromatogr A 794:147
17. Hogenboom AC, Hofman MP, Jolly DA, Niessen WMA, Brinkman UAT (2000) J Chromatogr A 885:377
18. Slobodnik J, Oztezkizan O, Lingeman H, Brinkman UAT (1996) J Chromatogr A 750:227
19. Slobodnik J, Ramalho S, Van Baar BLM, Louter AJH, Brinkman UAT (2000) Chemosphere 41:1469
20. Weller MG (2000) Fresenius J Anal Chem 366:635
21. Delaunay N, Pichon V, Hennion MC (2000) J Chromatogr B Biomed Sci Appl 745:15
22. Bean KA, Henion JD (1997) J Chromatogr A 791:119
23. Martin-Esteban A, Fernandez P, Stevenson D, Camara C (1997) Analyst 122:1113
24. Pichon V, Chen L, Hennion MC, Daniel R, Martel A, Le Goffic F, Abian J, Barcelo D (1995) Anal Chem 67:2451
25. Ferguson PL, Iden CR, McElroy AE, Brownawell BJ (2001) Anal Chem 73:3890
26. Rodriguez-Mozaz S, Lopez de Alda MJ, Barcelo D (2007) J Chromatogr A 1152:97
27. Deinl I, Angermaier L, Franzelius C, MacHbert G (1997) J Chromatogr B Biomed Appl 704:251
28. Nedved ML, Habibi-Goudarzi S, Ganem B, Henion JD (1996) Anal Chem 68:4228
29. Creaser CS, Feely SJ, Houghton E, Seymour M (1998) J Chromatogr A 794:37
30. Rhemrev-Boom MM, Yates M, Rudolph M, Raedts M (2001) J Pharm Biomed Anal 24:825
31. Delaunay-Bertoncini N, Hennion MC (2004) J Pharm Biomed Anal 34:717
32. Qiao F, Sun H, Yan H, Row KH (2006) Chromatographia 64:625
33. Pichon V (2007) J Chromatogr A 1152:41
34. Dong X, Wang N, Wang S, Zhang X, Fan Z (2004) J Chromatogr A 1057:13
35. Zhu X, Yang J, Su Q, Cai J, Gao Y (2005) J Chromatogr A 1092:161
36. Pap T, Horvath V, Tolokan A, Horvai G, Sellergren B (2002) J Chromatogr A 973:1
37. Turiel E, Martin-Esteban A, Fernandez P, Perez-Cond C, Camara C (2001) Anal Chem 73:5133
38. Watabe Y, Kubo T, Nishikawa T, Fujita T, Kaya K, Hosoya K (2006) J Chromatogr A 1120:252
39. Watabe Y, Kondo T, Morita M, Tanaka N, Haginaka J, Hosoya K (2004) J Chromatogr A 1032:45
40. Whitcombe MJ, Martin L, Vulfson EN (1998) Chromatographia 47:457
41. Dickert FL, Lieberzeit P, Tortschanoff M (2000) Sens Actuators B 65:186
42. Bolisay LD, Culver JN, Kofinas P (2006) Biomaterials 27:4165
43. Wei HS, Tsai YL, Wu JY, Chen H (2006) J Chromatogr B 836:57
44. Shea KJ, Sasaki DY (1989) J Am Chem Soc 111:3442
45. Rimmer S (1998) Chromatographia 47:470
46. Lavignac N, Allender CJ, Brain KR (2004) Anal Chim Acta 510:139
47. Vlatakis G, Andersson LI, Miller R, Mosbach K (1993) Nature 361:645

48. Baggiani C, Anfossi L, Baravalle P, Giovannoli C, Tozzi C (2005) Anal Chim Acta 531:199
49. Sellergren B, Shea KJ (1995) J Chromatogr A 690:29
50. Haginaka J, Kagawa C (2002) J Chromatogr A 948:77
51. Hosoya K, Yoshizako K, Shirasu Y, Kimata K, Araki T, Tanaka N, Haginaka J (1996) J Chromatogr A 728:139
52. Pang X, Cheng G, Li R, Lu S, Zhang Y (2005) Anal Chim Acta 550:13
53. Mayes AG, Mosbach K (1996) Anal Chem 68:3769
54. Downey JS, McIsaac G, Frank RS, Stöver HDH (2001) Macromolecules 34:4534
55. Ho KC, Yeh WM, Tung TS, Liao JY (2005) Anal Chim Acta 542:90
56. Venn RF, Goody RJ (1999) Chromatographia 50:407
57. Koeber R, Fleischer C, Lanza F, Boos KS, Sellergren B, Barceló D (2001) Anal Chem 73:2437
58. Lamprecht G, Kraushofer T, Stoschitzky K, Lindner W (2000) J Chromatogr B Biomed Sci Appl 740:219
59. El Mahjoub A, Staub C (2000) J Chromatogr B Biomed Sci Appl 742:381
60. Yu Z, Westerlund D, Boos KS (1997) J Chromatogr B Biomed Appl 704:53
61. Gorecki T, Namienik J (2002) TrAC-Trends Anal Chem 21:276
62. Vrana B, Allan IJ, Greenwood R, Mills GA, Dominiak E, Svensson K, Knutsson J, Morrison G (2005) TrAC-Trends Anal Chem 24:845
63. Koester CJ, Moulik A (2005) Anal Chem 77:3737
64. Koester CJ, Simonich SL, Esser BK (2003) Anal Chem 75:2813
65. Lord H, Pawliszyn J (2000) J Chromatogr A 885:153
66. Ouyang G, Pawliszyn J (2006) Anal Bioanal Chem 386:1059
67. Wu J, Yu X, Lord H, Pawliszyn J (2000) Analyst 125:391
68. Bruheim I, Liu X, Pawliszyn J (2003) Anal Chem 75:1002
69. Eisert R, Pawliszyn J (1997) Anal Chem 69:3140
70. Globig D, Weickhardt C (2005) Anal Bioanal Chem 381:656
71. Wu J, Tragas C, Lord H, Pawliszyn J (2002) J Chromatogr A 976:357
72. Gou Y, Eisert R, Pawliszyn J (2000) J Chromatogr A 873:137
73. Gou Y, Pawliszyn J (2000) Anal Chem 72:2774
74. Gou Y, Tragas C, Lord H, Pawliszyn J (2000) J Microcolumn Sep 12:125
75. Takino M, Daishima S, Nakahara T (2001) Analyst 126:602
76. Lee MR, Lee RJ, Lin YW, Chen CM, Hwang BH (1998) Anal Chem 70:1963
77. Rodriguez I, Rubi E, Gonzalez R, Quintana JB, Cela R (2005) Anal Chim Acta 537:259
78. Stashenko EE, Martinez JR (2004) TrAC-Trends Anal Chem 23:553
79. Dietz C, Sanz J, Camara C (2006) J Chromatogr A 1103:183
80. Camel V (2002) Anal Bioanal Chem 372:39
81. Covaci A, Voorspoels S, de Boer J (2003) Environ Int 29:735
82. Eljarrat E, Barcelo D (2004) TrAC-Trends Anal Chem 23:727
83. Eljarrat E, De La Cal A, Raldua D, Duran C, Barcelo D (2004) Environ Sci Technol 38:2603
84. De Voogt P, Kwast O, Hendriks R, Jonkers N (2000) Analysis 28:776
85. Zhao M, Van Der Wielen F, De Voogt P (1999) J Chromatogr A 837:129
86. Syage JA, Nies BJ, Evans MD, Hanold KA (2001) J Am Soc Mass Spectrom 12:648
87. Cochran JW (2002) J Chromatogr Sci 40:254
88. Hada M, Takino M, Yamagami T, Daishima S, Yamaguchi K (2000) J Chromatogr A 874:81
89. Santos FJ, Galceran MT (2002) TrAC-Trends Anal Chem 21:672
90. Gaines RB, Ledford EB Jr, Stuart JD (1998) J Microcolumn Sep 10:597

91. Hyotylainen T, Kallio M, Hartonen K, Jussila M, Palonen S, Riekkola ML (2002) Anal Chem 74:4441
92. Petrovic M, Gros M, Barcelo D (2007) In: Petrovic M, Barcelo D (eds) Comprehensive analytical chemistry. Elsevier, Amsterdam, p 157
93. Petrovic M, Gros M, Barcelo D (2006) J Chromatogr A 1124:68
94. Stuber M, Reemtsma T (2004) Anal Bioanal Chem 378:910
95. Alder L, Luderitz S, Lindtner K, Stan HJ (2004) J Chromatogr A 1058:67
96. Benijts T, Lambert W, De Leenheer A (2004) Anal Chem 76:704
97. Kloepfer A, Quintana JB, Reemtsma T (2005) J Chromatogr A 1067:153
98. Van De Steene JC, Mortier KA, Lambert WE (2006) J Chromatogr A 1123:71
99. Vanderford BJ, Snyder SA (2006) Environ Sci Technol 40:7312
100. Hopfgartner G, Husser C, Zell M (2003) J Mass Spectrom 38:138
101. Raffaelli A, Saba A (2003) Mass Spectrom Rev 22:318
102. Hanold KA, Fischer SM, Cormia PH, Miller CE, Syage JA (2004) Anal Chem 76:2842
103. Zwiener C, Frimmel FH (2004) Anal Bioanal Chem 378:851
104. Taniyasu S, Kannan K, Horii Y, Hanari N, Yamashita N (2003) Environ Sci Technol 37:2634
105. Villagrasa M, López de Alda MJ, Barceló D (2006) Anal Bioanal Chem 386:953
106. Schultz MM, Barofsky DF, Field JA (2006) Environ Sci Technol 40:289
107. Taniyasu S, Kannan K, Man KS, Gulkowska A, Sinclair E, Okazawa T, Yamashita N (2005) J Chromatogr A 1093:89
108. Karrman A, Van Bavel B, Järnberg U, Hardell L, Lindstrøm G (2005) Anal Chem 77:864
109. Yamashita N, Kannan K, Taniyasu S, Horii Y, Okazawa T, Petrick G, Gamo T (2004) Environ Sci Technol 38:5522
110. Takino M, Daishima S, Nakahara T (2003) Rapid Commun Mass Spectrom 17:383
111. Saito N, Sasaki K, Nakatome K, Harada K, Yoshinaga T, Koizumi A (2003) Arch Environ Contam Toxicol 45:149
112. Saito N, Harada K, Inoue K, Sasaki K, Yoshinaga T, Koizumi A (2004) J Occup Health 46:49
113. Pocurull E, Aguilar C, Alonso MC, Barcelo D, Borrull F, Marce RM (1999) J Chromatogr A 854:187
114. Higgins CP, Field JA, Criddle CS, Luthy RG (2005) Environ Sci Technol 39:3946
115. Schroder HF (2003) J Chromatogr A 1020:131
116. Moody CA, Field JA (1999) Environ Sci Technol 33:2800
117. Moody CA, Field JA (2000) Environ Sci Technol 34:3864
118. Ohya T, Kudo N, Suzuki E, Kawashima Y (1998) J Chromatogr B Biomed Appl 720:1
119. Abe T, Baba H, Itoh E, Tanaka K (2001) J Chromatogr A 920:173
120. Abe T, Baba H, Soloshonok I, Tanaka K (2000) J Chromatogr A 884:93
121. Hori H, Hayakawa E, Yamashita N, Taniyasu S, Nakata F, Kobayashi Y (2004) Chemosphere 57:273
122. Kuehl DW, Rozynov B (2003) Rapid Commun Mass Spectrom 17:2364
123. Kuklenyik Z, Reich JA, Tully JS, Needham LL, Calafat AM (2004) Environ Sci Technol 38:3698
124. Hansen KJ, Johnson HO, Eldridge JS, Butenhoff JL, Dick LA (2002) Environ Sci Technol 36:1681
125. Sumpter JP, Johnson AC (2005) Environ Sci Technol 39:4321
126. Kuster M, Lopez de Alda MJ, Barcelo D (2004) TrAC-Trends Anal Chem 23:790
127. Diaz-Cruz MS, Lopez de Alda MJ, Barcelo D (2003) TrAC-Trends Anal Chem 22:340
128. Petrovic M, Eljarrat E, Lopez de Alda MJ, Barcelo D (2002) J Chromatogr A 974:23

129. Lopez de Alda MJ, Barcelo D (2001) J Chromatogr A 938:145
130. Petrovic M, Eljarrat E, Lopez de Alda MJ, Barcelo D (2001) TrAC-Trends Anal Chem 20:637
131. Ying GG, Kookana RS, Ru YJ (2002) Environ Int 28:545
132. Hanselman TA, Graetz DA, Wilkie AC (2003) Environ Sci Technol 37:5471
133. Kuster M, Lopez de Alda M, Rodriguez-Mozaz S, Barcelo D (2007) In: Petrovic M, Barcelo D (eds) Comprehensive analytical chemistry. Elsevier, Amsterdam, p 219
134. Fine DD, Breidenbach GP, Price TL, Hutchins SR (2003) J Chromatogr A 1017:167
135. Liu R, Zhou JL, Wilding A (2004) J Chromatogr A 1022:179
136. Isobe T, Shiraishi H, Yasuda M, Shinoda A, Suzuki H, Morita M (2003) J Chromatogr A 984:195
137. Lopez de Alda MJ, Barcelo D (2001) J Chromatogr A 911:203
138. Rodriguez-Mozaz S, Lopez de Alda MJ, Barcelo D (2004) Anal Chem 76:6998
139. Tozzi C, Anfossi L, Giraudi G, Giovannoli C, Baggiani C, Vanni A (2002) J Chromatogr A 966:71
140. Penalver A, Pocurull E, Borrull F, Marce RM (2002) J Chromatogr A 964:153
141. Mitani K, Fujioka M, Kataoka H (2005) J Chromatogr A 1081:218
142. Braun P, Moeder M, Schrader S, Popp P, Kuschk P, Engewald W (2003) J Chromatogr A 988:41
143. Zang X, Luo R, Song N, Chen TK, Bozigian H (2005) Rapid Commun Mass Spectrom 19:3259
144. Kuster M, Lopez de Alda MJ, Barceló D (2005) Handbook of environmental chemistry, vol 2. Springer, Heidelberg
145. Petrovic M, Tavazzi S, Barcelo D (2002) J Chromatogr A 971:37
146. Cespedes R, Petrovic M, Raldua D, Saura U, Pina B, Lacorte S, Viana P, Barcelo D (2004) Anal Bioanal Chem 378:697
147. Liu R, Zhou JL, Wilding A (2004) J Chromatogr A 1038:19
148. Peck M, Gibson RW, Kortenkamp A, Hill EM (2004) Environ Toxicol Chem 23:945
149. Ternes TA, Andersen H, Gilberg D, Bonerz M (2002) Anal Chem 74:3498
150. Williams RJ, Johnson AC, Smith JJL, Kanda R (2003) Environ Sci Technol 37:1744
151. Peng X, Wang Z, Yang C, Chen F, Mai B (2006) J Chromatogr A 1116:51
152. Reddy S, Brownawell BJ (2005) Environ Toxicol Chem 24:1041
153. Lopez de Alda MJ, Gil A, Paz E, Barcelo D (2002) Analyst 127:1299
154. Ying GG, Kookana RS (2003) Environ Sci Technol 37:1256
155. Ying GG, Kookana RS, Dillon P (2003) Water Res 37:3785
156. Desbrow C, Routledge EJ, Brighty GC, Sumpter JP, Waldock M (1998) Environ Sci Technol 32:1549
157. Ingrand V, Herry G, Beausse J, De Roubin MR (2003) J Chromatogr A 1020:99
158. Larsson DGJ, Adolfsson-Erici M, Parkkonen J, Pettersson M, Berg AH, Olsson PE, Forlin L (1999) Aquat Toxicol 45:91
159. Belfroid AC, Van Der Horst A, Vethaak AD, Schafer AJ, Rijs GBJ, Wegener J, Cofino WP (1999) Sci Total Environ 225:101
160. Johnson AC, Belfroid A, Di Corcia A (2000) Sci Total Environ 256:163
161. Huang CH, Sedlak DL (2001) Environ Toxicol Chem 20:133
162. Rodgers-Gray TP, Jobling S, Morris S, Kelly C, Kirby S, Janbakhsh A, Harries JE, Waldock MJ, Sumpter JP, Tyler CR (2000) Environ Sci Technol 34:1521
163. Ternes TA, Stumpf M, Mueller J, Haberer K, Wilken RD, Servos M (1999) Sci Total Environ 225:81
164. Servos MR, Bennie DT, Burnison BK, Jurkovic A, McInnis R, Neheli T, Schnell A, Seto P, Smyth SA, Ternes TA (2005) Sci Total Environ 336:155

165. Kuch HM, Ballschmiter K (2000) Fresenius J Anal Chem 366:392
166. Beck IC, Bruhn R, Gandrass J, Ruck W (2005) J Chromatogr A 1090:98
167. Zuehlke S, Dunnbier U, Heberer T, Fritz B (2004) Ground Water Monit Rem 24:78
168. Zuehlke S, Duennbier U, Heberer T (2005) J Sep Sci 28:52
169. Quintana JB, Rodil R, Reemtsma T (2004) J Chromatogr A 1061:19
170. Gomes RL, Birkett JW, Scrimshaw MD, Lester JN (2005) Int J Environ Anal Chem 85:1
171. Shimada K, Mitamura K, Higashi T (2001) J Chromatogr A 935:141
172. Lopez de Alda MJ, Diaz-Cruz S, Petrovic M, Barcelo D (2003) J Chromatogr A 1000:503
173. Atkinson S, Atkinson MJ, Tarrant AM (2003) Environ Health Perspect 111:531
174. Schneider C, Scholer HF, Schneider RJ (2005) Anal Chim Acta 551:92
175. Hintemann T, Schneider C, Scholer HF, Schneider RJ (2006) Water Res 40:2287
176. Barel-Cohen K, Shore LS, Shemesh M, Wenzel A, Mueller J, Kronfeld-Schor N (2006) J Environ Manage 78:16
177. Soto AM, Maffini MV, Schaeberle CM, Sonnenschein C (2006) Best Pract Res Clin Endocrinol Metab 20:15
178. Clode SA (2006) Best Pract Res Clin Endocrinol Metab 20:35
179. Rodriguez-Mozaz S, Marco MP, Lopez de Alda MJ, Barcelo D (2004) Anal Bioanal Chem 378:588
180. Rodriguez-Mozaz S, Lopez de Alda MJ, Barcelo D (2006) Talanta 69:377
181. Nakamura S, Hwee Sian T, Daishima S (2001) J Chromatogr A 919:275
182. Cathum S, Sabik H (2001) Chromatographia 53:s-394
183. Xiao XY, McCalley DV, McEvoy J (2001) J Chromatogr A 923:195
184. Lerch O, Zinn P (2003) J Chromatogr A 991:77
185. Kuch HM, Ballschmiter K (2001) Environ Sci Technol 35:3201
186. Shareef A, Angove MJ, Wells JD (2006) J Chromatogr A 1108:121
187. Shareef A, Parnis CJ, Angove MJ, Wells JD, Johnson BB (2004) J Chromatogr A 1026:295
188. Labadie P, Budzinski H (2005) Environ Sci Technol 39:5113
189. Díaz-Cruz MS, Barceló D (2006) Anal Bioanal Chem 386:973
190. Gros M, Petrovic M, Barcelo D (2006) Anal Bioanal Chem 386:941
191. Diaz-Cruz MS, Barcelo D (2005) TrAC-Trends Anal Chem 24:645
192. Fatta D, Achilleos A, Nikolaou A, Meric S (2007) TrAC-Trends Anal Chem 26:515
193. Farre M, Petrovic M, Barcelo D (2007) Anal Bioanal Chem 387:1203
194. Jacobsen AM, Halling-Sørensen B, Ingerslev F, Hansen SH (2004) J Chromatogr A 1038:157
195. Loffler D, Ternes TA (2003) J Chromatogr A 1021:133
196. Schlusener MP, Spiteller M, Bester K (2003) J Chromatogr A 1003:21
197. Ternes TA, Bonerz M, Herrmann N, Loffler D, Keller E, Lacida BB, Alder AC (2005) J Chromatogr A 1067:213
198. Turiel E, Martin-Esteban A, Tadeo JL (2006) Anal Chim Acta 562:30
199. Hernandez F, Sancho JV, Ibanez M, Guerrero C (2007) TrAC-Trends Anal Chem 26:466
200. Gomez MJ, Petrovic M, Fernandez-Alba AR, Barcelo D (2006) J Chromatogr A 1114:224
201. Castiglioni S, Bagnati R, Calamari D, Fanelli R, Zuccato E (2005) J Chromatogr A 1092:206
202. Hao C, Lissemore L, Nguyen B, Kleywegt S, Yang P, Solomon K (2006) Anal Bioanal Chem 384:505

203. Stolker AAM, Niesing W, Hogendoorn EA, Versteegh JFM, Fuchs R, Brinkman UAT (2004) Anal Bioanal Chem 378:955
204. Kasprzyk-Hordern B, Dinsdale RM, Guwy AJ (2007) J Chromatogr A 1161:132
205. Nebot C, Gibb SW, Boyd KG (2007) Anal Chim Acta 598:87
206. Zhang ZL, Zhou JL (2007) J Chromatogr A 1154:205
207. Botitsi E, Frosyni C, Tsipi D (2007) Anal Bioanal Chem 387:1317
208. Trenholm RA, Vanderford BJ, Holady JC, Rexing DJ, Snyder SA (2006) Chemosphere 65:1990
209. Roberts PH, Bersuder P (2006) J Chromatogr A 1134:143
210. Naidong W, Roets E, Busson R, Hoogmartens J (1990) J Pharm Biomed Anal 8:881
211. Bryan PD, Hawkins KR, Stewart JT, Capomacchia AC (1992) Biomed Chromatogr 6:305
212. Gros M, Pizzolato TM, Petrovic M, Lopez de Alda MJ, Barcelo D (2007) J Chromatogr A, in press; doi:10.1016/jchroma.2007.10.052
213. Bravo JC, Garcinuno RM, Fernandez P, Durand JS (2007) Anal Bioanal Chem 388:1039
214. O'Connor S, Aga DS (2007) TrAC-Trends Anal Chem 26:456
215. Chapuis F, Mullot JU, Pichon V, Tuffal G, Hennion MC (2006) J Chromatogr A 1135:127
216. Kolpin DW, Furlong ET, Meyer MT, Thurman EM, Zaugg SD, Barber LB, Buxton HT (2002) Environ Sci Technol 36:1202
217. Metcalfe CD, Koenig BG, Bennie DT, Servos M, Ternes TA, Hirsch R (2003) Environ Toxicol Chem 22:2872
218. Weigel S, Berger U, Jensen E, Kallenborn R, Thoresen H, Huhnerfuss H (2004) Chemosphere 56:583
219. Bendz D, Paxéus NA, Ginn TR, Loge FJ (2005) J Hazard Mater 122:195
220. Perez S, Barcelo D (2007) Trends Anal Chem 26:494
221. Eichhorn P, Ferguson PL, Perez S, Aga DS (2005) Anal Chem 77:4176
222. Marchese S, Gentili A, Perret D, D'Ascenzo G, Pastori F (2003) Rapid Commun Mass Spectrom 17:879
223. Gomez MJ, Malato O, Ferrer I, Aguera A, Fernandez-Alba AR (2007) J Environ Monit 9:719
224. Pozo OJ, Guerrero C, Sancho JV, Ibanez M, Pitarch E, Hogendoorn E, Hernandez F (2006) J Chromatogr A 1103:83
225. Seitz W, Schulz W, Weber WH (2006) Rapid Commun Mass Spectrom 20:2281
226. Nikolai LN, McClure EL, MacLeod SL, Wong CS (2006) J Chromatogr A 1131:103
227. Quintana JB, Reemtsma T (2004) Rapid Commun Mass Spectrom 18:765
228. Peck AM (2006) Anal Bioanal Chem 386:907
229. Bester K, Huhnerfuss H, Lange W, Rimkus GG, Theobald N (1998) Water Res 32:1857
230. McAvoy DC, Schatowitz B, Jacob M, Hauk A, Eckhoff WS (2002) Environ Toxicol Chem 21:1323
231. Aguera A, Fernandez-Alba AR, Piedra L, Mezcua M, Gomez MJ (2003) Anal Chim Acta 480:193
232. Winkler M, Headley JV, Peru KM (2000) J Chromatogr A 903:203
233. Artola-Garicano E, Borkent I, Hermens JLM, Vaes WHJ (2003) Environ Sci Technol 37:3111
234. Ricking M, Schwarzbauer J, Hellou J, Svenson A, Zitko V (2003) Mar Pollut Bull 46:410
235. Dsikowitzky L, Schwarzbauer J, Littke R (2002) Org Geochem 33:1747
236. Winkler M, Kopf G, Hauptvogel C, Neu T (1998) Chemosphere 37:1139

237. Gatermann R, Huhnerfuss H, Rimkus G, Attar A, Kettrup A (1998) Chemosphere 36:2535
238. Gatermann R, Huhnerfuss H, Rimkus G, Wolf M, Franke S (1995) Mar Pollut Bull 30:221
239. Bester K (2005) Arch Environ Contam Toxicol 49:9
240. Bester K (2003) Water Res 37:3891
241. Lee HB, Peart TE, Sarafin K (2003) Water Qual Res J Canada 38:683
242. Paxeus N (2004) Water Sci Technol 50:253
243. Standley LJ, Kaplan LA, Smith D (2000) Environ Sci Technol 34:3124
244. Difrancesco AM, Chiu PC, Standley LJ, Allen HE, Salvito DT (2004) Environ Sci Technol 38:194
245. Simonich SL, Begley WM, Debaere G, Eckhoff WS (2000) Environ Sci Technol 34:959
246. Simonich SL, Federle TW, Eckhoff WS, Rottiers A, Webb S, Sabaliunas D, De Wolf W (2002) Environ Sci Technol 36:2839
247. Sakkas VA, Giokas DL, Lambropoulou DA, Albanis TA (2003) J Chromatogr A 1016:211
248. Lambropoulou DA, Giokas DL, Sakkas VA, Albanis TA, Karayannis MI (2002) J Chromatogr A 967:243
249. Osemwengie LI, Gerstenberger SL (2004) J Environ Monit 6:533
250. Osemwengie LI, Steinberg S (2001) J Chromatogr A 932:107
251. Lindstrom A, Buerge IJ, Poiger T, Bergqvist PA, Muller MD, Buser HR (2002) Environ Sci Technol 36:2322
252. Poiger T, Buser HR, Müller MD, Balmer ME, Buerge IJ (2003) Chimia 57:492
253. Buerge IJ, Buser HR, Müller MD, Poiger T (2003) Environ Sci Technol 37:5636
254. Peck AM, Hornbuckle KC (2004) Environ Sci Technol 38:367
255. Boyd GR, Palmeri JM, Zhang S, Grimm DA (2004) Sci Total Environ 333:137
256. Boyd GR, Reemtsma H, Grimm DA, Mitra S (2003) Sci Total Environ 311:135
257. Van Stee LLP, Leonards PEG, Van Loon WMGM, Hendriks AJ, Maas JL, Struijs J, Brinkman UAT (2002) Water Res 36:4455
258. Yang JJ, Metcalfe CD (2006) Sci Total Environ 363:149
259. Burkhardt MR, ReVello RC, Smith SG, Zaugg SD (2005) Anal Chim Acta 534:89
260. Zeng X, Sheng G, Xiong Y, Fu J (2005) Chemosphere 60:817
261. Stevens JL, Northcott GL, Stern GA, Tomy GT, Jones KC (2003) Environ Sci Technol 37:462
262. Morales-Munoz S, Luque-Garcia JL, Ramos MJ, Fernandez-Alba A, De Castro MDL (2005) Anal Chim Acta 552:50
263. Morales-Munoz S, Luque-Garcia JL, Ramos MJ, Martinez-Bueno MJ, De Castro MDL (2005) Chromatographia 62:69
264. Kupper T, Berset JD, Etter-Holzer R, Furrer R, Tarradellas J (2004) Chemosphere 54:1111
265. Herren D, Berset JD (2000) Chemosphere 40:565
266. Morales S, Canosa P, Rodriguez I, Rubi E, Cela R (2005) J Chromatogr A 1082:128
267. Felix T, Hall BJ, Brodbelt JS (1998) Anal Chim Acta 371:195
268. Llompart M, Garcia-Jares C, Salgado C, Polo M, Cela R (2003) J Chromatogr A 999:185
269. Gonzalez S, Barcelo D, Petrovic M (2007) TrAC-Trends Anal Chem 26:116
270. Lee HB (1999) Water Qual Res J Canada 34:3
271. Petrovic M, Barcelo D (2001) J Mass Spectrom 36:1173
272. Di Corcia A, Cavallo R, Crescenzi C, Nazzari M (2000) Environ Sci Technol 34:3914
273. Ahel M, Giger W, Koch M (1994) Water Res 28:1131

274. Di Corcia A, Costantino A, Crescenzi C, Marinoni E, Samperi R (1998) Environ Sci Technol 32:2401
275. Jonkers N, Knepper TP, De Voogt P (2001) Environ Sci Technol 35:335
276. Ventura F, Figueras A, Caixach J, Espadaler I, Romero J, Guardiola J, Rivera J (1988) Water Res 22:1211
277. Petrovic M, Diaz A, Ventura F, Barcelo D (2001) Anal Chem 73:5886
278. Petrovic M, Barcelo D (2000) Anal Chem 72:4560
279. Petrovic M, Fernandez-Alba AR, Borrull F, Marce RM, Mazo EG, Barcelo D (2002) Environ Toxicol Chem 21:37
280. Petrovic M, Lacorte S, Viana P, Barcelo D (2002) J Chromatogr A 959:15
281. Gonzalez S, Petrovic M, Barcelo D (2004) J Chromatogr A 1052:111
282. Petrovic M, Barcelo D (2004) TrAC-Trends Anal Chem 23:762
283. Gonzalez S, Petrovic M, Barcelo D (2007) Chemosphere 67:335
284. Suter MJF, Reiser R, Giger W (1996) J Mass Spectrom 31:357
285. Bennie DT, Sullivan CA, Lee HB, Peart TE, Maguire RJ (1997) Sci Total Environ 193:263
286. Lee HB, Peart TE (1995) Anal Chem 67:1976
287. Ding WH, Chen CT (1999) J Chromatogr A 862:113
288. Hao C, Croley TR, March RE, Koenig BG, Metcalfe CD (2000) J Mass Spectrom 35:818
289. Schroder HF (2001) J Chromatogr A 926:127
290. Houde F, DeBlois C, Berryman D (2002) J Chromatogr A 961:245
291. Petrovic M, Barcelo D, Diaz A, Ventura F (2003) J Am Soc Mass Spectrom 14:516
292. Ayorinde FO, Elhilo E (1999) Rapid Commun Mass Spectrom 13:2166
293. Ayorinde FO, Eribo BE, Johnson JH Jr, Elhilo E (1999) Rapid Commun Mass Spectrom 13:1124
294. Andreu V, Pico Y (2004) Anal Chem 76:2878
295. Cantero M, Rubio S, Perez-Bendito D (2006) J Chromatogr A 1120:260
296. Cantero M, Rubio S, Perez-Bendito D (2004) J Chromatogr A 1046:147
297. Hanton SD, Parees DM, Zweigenbaum J (2006) J Am Soc Mass Spectrom 17:453
298. Willetts M, Clench MR, Greenwood R, Mills G, Carolan V (1999) Rapid Commun Mass Spectrom 13:251
299. Covaci A, Voorspoels S, Ramos L, Neels H, Blust R (2007) J Chromatogr A 1153:145
300. Luo Q, Cai ZW, Wong MH (2007) Sci Total Environ 383:115
301. Xiang CH, Luo XJ, Chen SJ, Yu M, Mai BX, Zeng EY (2007) Environ Toxicol Chem 26:616
302. Labandeira A, Eljarrat E, Barcelo D (2007) Environ Pollut 146:188
303. Streets SS, Henderson SA, Stoner AD, Carlson DL, Simcik MF, Swackhamer DL (2006) Environ Sci Technol 40:7263
304. Law K, Halldorson T, Danell R, Stern G, Gewurtz S, Alaee M, Marvin C, Whittle M, Tomy G (2006) Environ Toxicol Chem 25:2177
305. Gama AC, Sanatcumar P, Viana P, Barcelo D, Bordado JC (2006) Chemosphere 64:306
306. Eljarrat E, De La Cal A, Raldua D, Duran C, Barcelo D (2005) Environ Pollut 133:501
307. Hites RA (2004) Environ Sci Technol 38:945
308. Weber H, Heseker H (2004) Fresenius Environ Bull 13:356
309. Schecter A, Pavuk M, Papke O, Ryan JJ, Birnbaum L, Rosen R (2003) Environ Health Perspect 111:1723
310. Ikonomou MG, Rayne S, Addison RF (2002) Environ Sci Technol 36:1886
311. Guillamon M, Martinez E, Eljarrat E, Lacorte S (2002) Organohalogenated Compounds 55:199

312. Helleday T, Tuominen KL, Bergman A, Jenssen D (1999) Mutat Res Genet Toxicol Environ Mutagen 439:137
313. Meerts IATM, Letcher RJ, Hoving S, Marsh G, Bergman A, Lemmen JG, Van Der Burg B, Brouwer A (2001) Environ Health Perspect 109:399
314. Stapleton HM, Keller JM, Schantz MM, Kucklick JR, Leigh SD, Wise SA (2007) Anal Bioanal Chem 387:2365
315. Darnerud PO, Atuma S, Aune M, Cnattingus S, Wenroth ML, Wicklund-Glynn A (1998) Organohalogenated Compounds 35:411
316. Ohta S, Ishizuka D, Nishimura H, Nakao T, Aozasa O, Shimidzu Y, Ochiai F, Kida T, Nishi M, Miyata H (2002) Chemosphere 46:689
317. Booij K, Zegers BN, Boon JP (2002) Chemosphere 46:683
318. Hovander L, Malmberg T, Athanasiadou M, Athanassiadis I, Rahm S, Bergman A, Wehler EK (2002) Arch Environ Contam Toxicol 42:105
319. Sjodin A, Hagmar L, Klasson-Wehler E, Kronholm-Dlab K, Jakobsson E, Bergman A (1999) Environ Health Perspect 107:643
320. Fontanals N, Barri T, Bergstrom S, Jonsson JA (2006) J Chromatogr A 1133:41
321. Stapleton HM, Harner T, Shoeib M, Keller JM, Schantz MM, Leigh SD, Wise SA (2006) Anal Bioanal Chem 384:791
322. de Boer J, Allchin CR, Law R, Zegers BN, Boon JP (2001) Trends Anal Chem 20:591
323. De Boer J, Wester PG, van den Horst A, Leonards PEG (2003) Environ Pollut 122:63
324. Nylund K, Asplund L, Jansson B, Jonsson P, Litzen K, Sellstrom U (1992) Chemosphere 24:1721
325. Hartonen K, Bowadt S, Hawthorne SB, Riekkola ML (1997) J Chromatogr A 774:229
326. De La Cal A, Eljarrat E, Barcelo D (2003) J Chromatogr A 1021:165
327. Samara F, Tsai CW, Aga DS (2006) Environ Pollut 139:489
328. Yusa M, Pardo O, Pastro A, de la Guardia M (2006) Anal Chim Acta 557:304
329. Law RJ, Allchin CR, Bennett ME, Morris S, Rogan E (2002) Chemosphere 46:673
330. Bjorklund J, Tollback P, Hiarne C, Dyremark E, Ostman C (2004) J Chromatogr A 1041:201
331. Eljarrat E, De la Cal A, Barcelo D (2003) J Chromatogr A 1008:181
332. Alaee M, Backus S, Cannon C (2001) J Sep Sci 24:465
333. Zhu LY, Hites RA (2002) Environ Sci Technol 38:2779
334. Hale RC, La Guardia MJ, Harvey E, Gaylor MO, Mainor TM (2006) Chemosphere 64:181
335. Wise SA, Poster DL, Schantz MM, Kucklick JR, Sander LC, Lopez De Alda M, Schubert P, Parris RM, Porter BJ (2004) Anal Bioanal Chem 378:1251
336. Korytar P, Covaci A, Leonards PEG, De Boer J, Brinkman UAT (2005) J Chromatogr A 1100:200
337. Alaee M, Sergeant DB, Ikonomou MG, Luross JM (2001) Chemosphere 44:1489
338. Polo M, Gomez-Noya G, Quintana JB, Llompart M, Garcia-Jares C, Cela R (2004) Anal Chem 76:1054
339. Wang D, Cai Z, Jiang G, Wong MH, Wong WK (2005) Rapid Commun Mass Spectrom 19:83
340. Focant JF, Sjodin A, Patterson DG Jr (2003) J Chromatogr A 1019:143
341. Cajka T, Hajslova J, Kazda R, Poustka J (2005) J Sep Sci 28:601
342. Debrauwer L, Riu A, Jouahri M, Rathahao E, Jouanin I, Antignac JP, Cariou R, Le Bizec B, Zalko D (2005) J Chromatogr A 1082:98
343. Hua W, Bennett ER, Letcher RJ (2005) Environ Int 31:621
344. Stapleton HM (2006) Anal Bioanal Chem 386:807

345. Johnson R, Pankow J, Bender D, Price C, Zogorski J (2000) Environ Sci Technol 34:210A
346. Squillace PJ, Pankow JF, Korte NE, Zogorski JS (1997) Environ Toxicol Chem 16:1836
347. Rosell M, Lacorte S, Ginebreda A, Barcelo D (2003) J Chromatogr A 995:171
348. Rosell M, Lacorte S, Barcelo D (2006) TrAC-Trends Anal Chem 25:1016
349. Atienza J, Aragon P, Herrero MA, Puchades R, Maquieira A (2005) Crit Rev Anal Chem 35:317
350. Rosell M, Lacorte S, Barcelo D (2006) J Chromatogr A 1132:28
351. Achten C, Kolb A, Puttmann W, Seel P, Gihr R (2002) Environ Sci Technol 36:3652
352. Schmidt TC, Duong HA, Berg M, Haderlein SB (2001) Analyst 126:405
353. Schmidt TC (2003) TrAC-Trends Anal Chem 22:776
354. Atienza J, Aragon P, Herrero MA, Puchades R, Maquieira A (2005) Crit Rev Anal Chem 35:317
355. Schuhmacher R, Fuhrer M, Kandler W, Stadlmann C, Krska R (2003) Anal Bioanal Chem 377:1140
356. Bellar T (1991) US Environmental Protection Agency, Environmental Monitoring Systems Laboratory, Cincinnati
357. Bianchi A, Varney MS (1989) J High Resolut Chromatogr 12:184
358. Bianchi AP, Varney MS, Phillips J (1991) J Chromatogr 542:413
359. Amaral OC, Olivella L, Grimalt JO, Albaiges J (1994) J Chromatogr A 675:177
360. Zuloaga O, Etxebarria N, Fernandez LA, Madariaga JM (2000) Anal Chim Acta 416:43
361. Campillo N, Vinas P, Lopez-Garcia I, Aguinaga N, Hernandez-Cordoba M (2004) Talanta 64:584
362. Tanabe A, Tsuchida Y, Ibaraki T, Kawata K, Yasuhara A, Shibamoto T (2005) J Chromatogr A 1066:159
363. Boulanger B, Vargo JD, Schnoor JL, Hornbuckle KC (2005) Environ Sci Technol 39:5524
364. Lagana A, Fago G, Marino A, Santarelli D (2001) Anal Lett 34:913
365. Stolker AAM, Niesing W, Fuchs R, Vreeken RJ, Niessen WMA, Brinkman UAT (2004) Anal Bioanal Chem 378:1754
366. Yang S, Cha J, Carlson K (2005) J Chromatogr A 1097:40
367. Montes R, Rodriguez I, Rubi E, Cela R (2007) J Chromatogr A 1143:41
368. Suzuki S, Hasegawa A (2006) Anal Sci 22:469
369. Martinez-Carballo E, Gonzalez-Barreiro C, Scharf S, Gans O (2007) Environ Pollut 148:570
370. Berset JD, Bigler P, Herren D (2000) Anal Chem 72:2124
371. Lee HB, Sarafin K, Peart TE, Svoboda ML (2003) Water Qual Res J Canada 38:667
372. Morris S, Allchin CR, Zegers BN, Haftka JJH, Boon JP, Belpaire C, Leonards PEG, Van Leeuwen SPJ, De Boer J (2004) Environ Sci Technol 38:5497
373. Fabrellas B, Sanz P, Larrazabal D, Abad E (2000) Organohalogenated Compounds 45:160
374. Stapleton HM, Brazil B, Holbrook RD, Mitchelmore CL, Benedict R, Konstantinov A, Potter D (2006) Environ Sci Technol 40:4653
375. Leon VM, Gonzalez-Mazo E, Gomez-Parra A (2000) J Chromatogr A 889:211
376. Voogt P, Saez M (2006) Trends Anal Chem 25:326
377. Ojala M, Mattila I, Tarkiainen V, Sarme T, Ketola RA, Maattanen A, Kosiainen R, Kotiaho T (2001) Anal Chem 73:3624

Hdb Env Chem Vol. 5, Part S/1 (2008): 105–142
DOI 10.1007/698_5_105

Published online: 21 March 2008

Acute and Chronic Effects of Emerging Contaminants

Tvrtko Smital

Laboratory for Molecular Ecotoxicology,
Division for Marine and Environmental Research,
Rudjer Boskovic Institute, Bijenicka 54, 10000 Zagreb, Croatia
smital@irb.hr

Abstract Acute or chronic toxicity profiling represents one of the critical elements for scientifically reliable characterization and prioritization of potentially hazardous contaminants. The very same is true for so-called emerging contaminants, regardless of the definition used in defining various aspects of "emerging", including substances that belong to new chemical classes, new types of use, new effects, mechanism of action, source, or exposure route. From the (eco)toxicological perspective, however, there are two essential drawbacks which prevent efficient characterization of risk posed to humans and the environment by the presence of emerging contaminants. First is related to the fact that the potential of analytical chemistry to measure contaminants currently exceeds our understanding of their potential environmental effects. Secondly, for most emerging contaminants there is currently little information regarding their potential toxicological significance in ecosystems, particularly the effects from long-term low-level environmental exposures. Based on these facts a brief overview of acute and chronic toxic effects on human and wildlife, reported for various classes of emerging contaminants, is presented in this chapter. The most demanding research unknowns, methodological drawbacks, and priorities will be highlighted, and finally, future strategies needed for efficient (eco)toxicological characterization of emerging contaminants will be suggested.

Keywords Acute and chronic toxicity · (Eco)toxicological characterization · Emerging contaminants

Abbreviations

AFOs	Animal feeding operations
ALS	Amyotrophic lateral sclerosis
ARGs	Antibiotic resistance genes
BADGE	Bisphenol A diglycidyl ether
BPA	Bisphenol A
CHE	The Collaborative on Health and the Environment
DES	Diethylstilbestrol
ELS	Early life-stages
GDS	Genotoxic disease syndrome
HAdV	Human adenoviruses
HEV	Hepatitis E virus
HPV	High Production Volume
MATC	Maximum acceptable toxicant concentration
MXR	Multixenobiotic resistance
NOAA	US National Centers for Coastal Ocean Science
OSPAR	Oslo and Paris Convention for the Protection of the Marine Environment of the North-East Atlantic
PBDEs	Polybrominated diphenyl ethers
PCNs	Polychlorinated naphthalenes
PCPs	Personal care products
PFCs	Perfluorochemicals
POPs	Persistent organic pollutants
PVC	Poly-vinyl chloride
QDs	Quantum dots
REACH	Registration, Evaluation, and Authorization of Chemicals
STP	Sewage treatment plant
US FDA	US Food and Drug Administration
USCDC	US Centers for Disease Control
USEPA	US Environmental Protection Agency
WWF	World Wide Fund

1
Introduction

Cancer, reproductive disorders, impaired neurological development, allergies – these are the types of health effects that make headlines. That puts corresponding chemicals "culprits" on the top of any list of emerging contaminants: potentially toxic substances whose effects or presence are poorly known, often because these chemicals have only begun to enter the human water or food supply. On the other hand, humans and wildlife are constantly exposed to a variety of contaminants present at low levels. These include both new chemicals, with previously unknown effects and those with well known acute (short-term exposure) human and ecological health effects. The result has been new research on emerging contaminants and an increased emphasis on methods of analyzing health effects of contaminants. The area in which several advances have recently been made is related to long-term health effects of chemical exposure. Other studies are now examining the impacts of organic compounds which may interfere with the endocrine systems of living organisms. Another active area of research is focused on how chemicals interact with each other and the natural environment. Finally, researchers are continuing to find new chemicals that bioaccumulate in the food chain. Such chemicals can be present in water at very low levels, however, they accumulate to higher concentrations in living tissue, substantially magnifying any health effects.

Three components have been usually considered to be critical for a chemical to be classified as highly hazardous contaminant: (1) persistence (structural stability resulting in long environmental half-lives); (2) lypophilicity (resulting in bioconcentration and possible biomagnification in the food chain); and (3) proven acute or chronic toxicity. However, all of these criteria need certain reconsideration – for example, continual release of some contaminants by the sewage treatment plants (STPs) give them a "pseudo-persistance" in aquatic environments; some drugs are actively transported in cells regardless of their lipid-water partition coefficients; finally, chemicals may act as indirect toxicants (such as nanoparticles or antibiotics, for example). Nevertheless, toxicity remains one of the cornerstones for scientifically reliable classification and hazard prioritization. From the (eco)toxicological perspective, however, two serious drawbacks appears to be essential in preventing efficient and reliable characterization of risk posed to humans and the environment by the presence of emerging contaminants.

Firstly, due to recent improvements in analytical chemistry, the types of chemicals that can be detected are increasing, and the limits of concentration at which they can be detected are continuously lowered. Our ability to measure contaminants currently exceeds our understanding of their potential environmental effects. Proving the link between real environmental exposure levels and acute or chronic toxic effects to humans and/or wildlife is an expensive, time-consuming, and complex research endeavor. Evaluat-

toxicological effects. Alternatively, new observations or information (e.g., endocrine disruption) on contaminants (e.g., nonylphenol) can cause the reconsideration of a well known contaminant as a (re)emerging contaminant. Unfortunately, the same analytical advances which bring contaminants to the public's attention do not offer knowledge about whether the newly detected contaminant is of (eco)toxicological interest. Assessing the effects of these contaminants in the environment remains a major time- and resource-intensive challenge. Therefore, it is not surprising that, for the many thousands of chemicals being produced or already on the market and the many new microbes that are being discovered, advances in our understanding of their (eco)toxicological properties are considerably slow and lag significantly behind the public's demand for information. As a result, a contaminant may be considered for several years to be emerging. Regardless of the definition in this chapter we will cover different dimensions of "emerging", including substances that belong to new chemical classes, new types of use, new effects, mechanism of action, source, or exposure route.

3 Human vs. Ecological Health Effects

3.1 Human Health Effects – Basic Principals

Human health results from complex interactions among genes and the environment. Environmental exposures to chemical, physical, and biological agents may cause or contribute to disease in susceptible individuals. Personal lifestyle factors, such as diet, smoking, alcohol use, level of exercise, and UV exposure, often are a primary focus when considering preventable causes of disease. However, exposures to chemical contaminants at work, at home, in the outdoors, and even in utero, are increasingly recognized as important and preventable contributors to human disease [3].

Toxic effects of chemical agents are often not well understood or appreciated by healthcare providers and the general public. Some chemicals, such as asbestos, vinyl chloride, and lead, are well established as causes of human disease. There is also good evidence available to suggest increases in the incidence of some cancers, asthma, and developmental disorders, can be attributed to chemical exposure, particularly in young children. Other diseases, such as amyotrophic lateral sclerosis (ALS) or Gulf War Syndrome have been hypothesized to be associated with chemical exposures, but the evidence is limited.

The effects of chemical exposures in humans are difficult to study, because controlled human experimentation is not ethically feasible. There is limited human data obtained from accidental exposures, overdoses, or studies of workers exposed occupationally. Environmental exposure studies in the general population also can be useful, though they often have limitations. Many diseases, such as cancer, may not appear until decades after an exposure has occurred, making it difficult for causal associations to be identified. Exposure assessment, a critical step in environmental epidemiologic studies, is difficult. Retrospective exposure assessment usually requires estimates and considerable judgment and is subject to significant error. An individual's exposure may change over time, and exposures to multiple chemicals occur both in the home and work environments. It is difficult for individuals to remember or even know what they have been exposed to. Furthermore, the effects of chemical exposures may vary, depending on the age of exposure (in utero, childhood, adult), the route of exposure (ingestion, inhalation, dermal), amount and duration of exposure, exposures to multiple chemicals simultaneously, and other personal susceptibility factors, including genetic variability.

Because of these challenges, most toxicity research is conducted in animal studies, which contribute important toxicological information and provide strong evidence of disease without human epidemiological studies if the mechanism of action is relevant. Many regulatory decisions to limit or ban the use of a chemical are based on animal data. Furthermore, human epidemiology studies are often conducted after an association has been hypothesized based on animal data. The same is true for most data related to human toxic effects of emerging contaminants described in this chapter.

Although there is a need for much more chemicals to be adequately characterized, a vast amount of data for human acute or chronic toxic effects of various contaminants is already available and published. What is often lacking, both for scientists and regulators, as well as for citizens, is a comprehensive and reliable tool that offers free, scientifically sound, and reliable information about contaminants hazardous to humans. Nevertheless, useful and comprehensive evidence has been recently complied within two independent sources. With the motto: "Mapping the Pollution in People", The Human Toxome Project at the Environmental Working Group in the USA [4] established a web database aimed at collecting and presenting relevant data about health effects of virtually all pollutants that enter the human body. Another source is The Collaborative on Health and the Environment (CHE) Toxicant and Disease Database [5], a searchable database that summarizes links between chemical contaminants and approximately 180 human diseases or conditions.

3.2 Ecotoxicological Aspects of Emerging Contaminants

As much as it is difficult to establish clear causal connections between contaminant(s) exposure and human health effects, it is far more difficult to do the same on the ecosystem level, with numerous species involved at different levels of biological organization, and many environmental factors that make the interpretation of field data even more complex. Paradoxically (or not?), knowledge, expertise, and resources being invested in human health issues, outmatch multiple times those invested in the environmental health arena, explaining to a large extent the critical shortage in data needed for a sustainable management of environmental resources.

More specifically, the objective of aquatic toxicity tests with effluents or pure compounds is to estimate the "safe" or "no effect" concentration of these substances, which is defined as the concentration that will permit normal propagation of fish and other aquatic life in the receiving waters. The endpoints which have been considered in tests to determine the adverse effects of toxicants include death and survival, decreased reproduction and growth, locomotor activity, gill ventilation rate, heart rate, blood chemistry, histopathology, enzyme activity, olfactory function, etc. [6]. Since it is not feasible to detect and/or measure all of these (and other possible) effects of toxic substances on a routine basis, observations in toxicity tests generally have been limited to only a few effects, typically including mortality, growth, and reproduction.

Acute lethality is an obvious and easily observed effect which accounts for its wide use in the early period of evaluation of the toxicity of pure compounds and complex effluents. The results of these tests were usually expressed as the concentration lethal to 50% of the test organisms (LC50) over relatively short exposure periods (one-to-four days).

As exposure periods of acute tests were lengthened, the LC50 and lethal threshold concentration were observed to decline for many compounds. By lengthening the tests to include one or more complete life cycles and observing the more subtle effects of the toxicants, such as a reduction in growth and reproduction, more accurate direct estimates of the threshold or safe concentration of the toxicant could be obtained. However, laboratory life-cycle tests may not accurately estimate the "safe" concentration of toxicants, because they are conducted with a limited number of species under highly controlled, steady-state conditions, and the results do not include the effects of the stresses to which the organisms would ordinarily be exposed in the natural environment.

An early published account of a full life-cycle fish toxicity test was that of Mount and Stephan back in 1967 [7]. In this study, fathead minnows, *Pimephales promelas*, were exposed to a graded series of pesticide concentrations throughout their life-cycle, and the effects of the toxicant on survival, growth, and reproduction were measured and evaluated. This work was soon followed by full life-cycle tests using other toxicants and fish species. McKim [8] evaluated the data from 56 full life-cycle tests, 32 of which used the fathead minnow, and concluded that the embryo-larval and early juvenile life-stages were the most sensitive stages. He proposed the use of partial life-cycle toxicity tests with the early life-stages (ELS) of fish to establish water quality criteria. Macek and Sleight [9] found that exposure of critical life-stages of fish to toxicants provides estimates of chronically safe concentrations remarkably similar to those derived from full life-cycle toxicity tests. They reported that for a great majority of toxicants, the concentration which will not be acutely toxic to the most sensitive life stages is the chronically safe concentration for fish, and that the most sensitive life stages are the embryos and fry. Critical life-stage exposure was considered to be exposure of the embryos during most, preferably all, of the embryogenic (incubation) period, and exposure of the fry for 30 days post-hatch for warm water fish with embryogenic periods ranging from 1–14 days, and for 60 days post-hatch for fish with longer embryogenic periods. They concluded that in the majority of cases, the maximum acceptable toxicant concentration (MATC) could be estimated from the results of exposure of the embryos during incubation, and the larvae for 30 days post-hatch.

In a review of the literature on 173 fish full life-cycle and ELS tests performed to determine the chronically safe concentrations of a wide variety of toxicants, such as metals, pesticides, organics, inorganics, detergents, and complex effluents, Woltering [10] found that at the lowest effect concentration, significant reductions were observed in fry survival in 57%, fry growth in 36%, and egg hatchability in 19% of the tests. He also found that fry survival and growth were often equally sensitive, and concluded that the growth response could be deleted from routine application of the ELS tests. The net result would be a significant reduction in the duration and cost of screening tests with no appreciable impact on estimating MATCs for chemical hazard assessments.

Efforts to further reduce the length of partial life-cycle toxicity tests for fish without compromising their predictive value have resulted in the development of an eight-day embryo-larval survival and teratogenicity test for fish and other aquatic vertebrates [11, 12], and a seven-day larval survival and growth test [13]. The similarity of estimates of chronically safe concentrations of toxicants derived from short-term embryo-larval survival and teratogenicity tests to those derived from full life-cycle tests has been firstly demonstrated by Birge et al. [12, 14].

Since that time, most of our knowledge about acute and chronic effects of contaminants originates from the described type of ecotoxicity tests. An overview of the present knowledge related to emerging contaminants/concerns will be presented in the next section.

4
Human and Environmental Health Effects

Among many different categories of emerging contaminants, we will especially take into consideration those which, according to the state-of-the-art litera-

Table 2 Major human/environmental health concerns and priority status of the most prominent categories of emerging contaminants

Health concern	Chemical family: Alkyl-phenols	Bisphenol A & BADGE	Brominated dioxins & furans	Per-chlor-ate	Perfluoro-chemicals (PFCs)	Phtal-ates	Polybrom-inated di-phenyl ethers (PBDEs)
Birth defects and developmental delays	+	+		+	+	+++	++
Brain and nervous system					++	+++	+++
Cancer		+	+		+	+	+
Endocrine system	+			+++	+		
Gastrointestinal (including liver)					+		+
Hematologic (blood) system				+			
Hormone activity	+	+++			+++	+++	+++
Immune system (including sensi-tization and allergies)		++	+		+++	+++	
Kidney and renal system		+			+++		
Reproduction and fertility	+++	++	++		+++	+++	+++
Skin		+				+	
Respiratory system	+					+++	
Wildlife and environ-mental toxicity	+++	++					+
Persistent, accumulates in wildlife and/or people	++	++	+++		++	++	++
OSPAR list	√	√				√	√
Priority substance and/ or banned in the EU, USA or Canada	√					√	

Weight of evidence: + limited; ++ probable; +++ strong

ture evidence, appear to be of the highest (eco)toxicological relevance and are frequently detected in industrial and/or municipal waste: industrial chemicals (new and recently recognized), personal care products, pharmaceuticals, nonculturable biological pathogens, and, finally, nanomaterials. Instead of referring to numerous studies utilizing various in vivo and in vitro test systems in attempts to characterize toxicity of many different contaminants, what follows in the section(s) below is a brief summary describing relevance and toxic effects reported with a reasonable weight of evidence for the most prominent emerging contaminants. Basic info referring to major human health concerns, wildlife toxicity, bioaccumulation/persistency potential, and the regulatory status of those substances is presented in Table 2.

Table 2 (continued)

Health concern	Chemical family Polychlorinated naphthalenes (PCNs)	Fragrances (nitro- and polycyclic musks)	Triclosan	Pharmaceuticals	Non-culturable biological pathogens	Nanomaterials
Birth defects and developmental delays				+++		
Brain and nervous system				+		+
Cancer		+		+		
Endocrine system		+	+	+++		
Gastrointestinal (including liver)	+++	+++		+	+++	
Hematologic (blood) system						
Hormone activity				+++		
Immune system (including sensitization and allergies)		+	+		+	+
Kidney and renal system						
Reproduction and fertility	+	+++	+	++		
Skin	+++	+	+	+	++	
Respiratory system				+	+++	++
Wildlife and environmental toxicity	++	+	+	++		+
Persistent, accumulates in wildlife and/or people	++	++	++	+		
OSPAR list	√	√				
Priority substance and/ or banned in the EU, USA or Canada	√		√			

Weight of evidence: + limited; ++ probable; +++ strong

4.1 Industrial Chemicals

4.1.1 Alkylphenols

Alkylphenols are widely used industrial chemicals which act as detergents or surfactants. They are added to cosmetics, paints, pesticides, detergents, and cleaning products. Alkylphenols have been recently detected in surface waters contaminated with urban runoff and in wastewater effluents [15, 16] and have been measured in air samples. One study found that newer homes, especially those with poly-vinyl chloride (PVC) materials, have more alkylphenol residues than older houses or outdoor air [17]. As a group they are highly toxic to aquatic organisms. Dozens of recent studies have documented the in vitro and in vivo estrogenic activity of alkylphenols in human cell lines and animals [18–20]. Recent study by McClusky and colleagues [21] revealed harmful effects of *p*-nonylphenol exposure to spermatogenic cycle in male rats. Similar estrogenic activities of alkylphenols have been reported for aquatic organisms, including a recent example of the reduction of reproductive competence of male fathead minnow upon exposure to environmentally relevant mixtures of alkylphenolethoxylates [22]. Further supported by their persistency in aquatic environments and bioaccumulation potential, alkylphenols are put on the OSPAR list of possible substances of concern and included in the list of priority substances in the EU water policy.

4.1.2 Bisphenol A and Bisphenol A Diglycidyl Ether

In use since the 1950's, bisphenol A (BPA) is a building block for polycarbonate plastic and epoxy resins. BPA and its derivative, bisphenol A diglycidyl ether (BADGE), are found in many everyday products, such as the lining of metal food and drink cans, plastic baby bottles, pacifiers, and baby toys, dental sealants, computers, cell phones, hard plastic water bottles (such as Nalgene), paints, adhesives, enamels, varnishes, CDs and DVDs, and certain microwavable or reusable food and drink containers. These compounds have been shown to leach into food and water from containers – particularly after heating or as plastic ages.

BPA is a hormone-mimicking chemical that can disrupt the endocrine system at very low concentrations. More than a hundred animal studies have linked low doses of bisphenol A to a variety of adverse health effects, such as reduced sperm count, impaired immune system functioning, increases in prostate tumor proliferation, altered prostate and uterus development, insulin resistance, alteration of brain chemistry, early puberty, and behavioral changes [23–36]. Significantly, many of the studies showing adverse effects are at levels many times lower than what the US Environmental Protection Agency (USEPA) considers safe (50 μg/kg/day).

For BADGE, a bisphenol A derivative used to make epoxy resins and in a variety of industrial, engineering, and construction applications, the major pathway for human exposure is through chemical leaching from the linings of food and drink cans. BADGE is also found in some dental sealants [37].

Some basic toxicological testing has been done on BADGE, but the compound has not been extensively studied. One of the most important toxicological questions is whether BADGE breaks down into bisphenol A in the human body. Based on urinary levels of BPA in workers exposed to BADGE versus unexposed controls, researchers concluded that BADGE breaks down into BPA in the body [38]. However, other research has suggested that there is no such biotransformation [39]. In the human body, BADGE appears in a hydrolysis product known as BADGE 40-H [40]. BADGE is quickly metabolized by the body (within a day or so), therefore body burden levels represent recent exposures.

Considering that its sister chemical, bisphenol A, has a non-monotonic dose response curve, showing nonintuitive patterns of toxicity, it would be difficult to make a final assessment on the toxicity of BADGE without more detailed study. There is some evidence that BADGE is a rodent carcinogen, but data for humans is lacking [41, 42]. Workers using epoxy resin in the construction industry have shown BADGE to be a contact allergen [43]. Males exposed to BADGE through spraying epoxy resin have associated depressed gonadotrophic hormones [38]. A study of BADGE given to pregnant rabbits found that at the lowest dose tested (30 mg/kg/day for days 6 to 18 of gestation) BADGE affected pregnancy ability and the sex ratio of their litters [39]. An in vitro study found that BADGE can induce time and dose-dependent morphological changes and cell detachment from the substratum and can inhibit cell proliferation [44]. Another study found that a BADGE derivative (BADGE.2HCl) can act as an androgen antagonist in in vitro systems [45].

4.1.3
Brominated Dioxins and Furans

Brominated dioxins and furans are toxic, persistent, bioaccumulative, and lipophilic ("fat-loving"). Along with dioxins, furans are pollutants produced during PVC plastic production, industrial bleaching, and incineration. They build up in human tissues, are stored in fatty tissues and fluid, such as breast milk, and can be passed on to fetuses and infants during pregnancy and lactation. Brominated dioxins and furans are formed unintentionally, either from incineration of wastes which include consumer products infused with brominated flame retardants, such as polybrominated diphenyl ethers (PBDEs), or as trace contaminants in mixtures of bromine-containing chemicals. Primary (eco)toxicological concern for brominated dioxins and furans is their dioxin-like activity, meaning that they cause birth defects in animals and otherwise disrupt reproductive development and the immune and hormone systems [46–49]. They add to the total dioxin body burden of people, which are near levels where adverse health effects may be occurring in the general population [50].

4.1.4
Perchlorate

The vast majority of perchlorate manufactured is used to make solid rocket and missile fuel, while smaller amounts of perchlorate are also used to make firework and road flares. Perchlorate is also a contaminant of certain types of fertilizer which were widely used in the early part of the 20th century, but are in limited use today [51]. According to the analysis of the USEPA's latest data, perchlorate is known to be contaminating at least 160 public drinking water systems in 26 US states [52]. Tests of almost 3000 human urine and breast milk samples, along with tests of more than 1000 fruit, vegetable, cow's milk, beer, and wine samples, reveal that perchlorate exposure in the population is pervasive. Every urine sample tested showed some level of perchlorate contamination, and almost 70% of the fruit and beverage samples tested have had detectable perchlorate [52–60].

Critical toxic effect of perchlorate is inhibition of the thyroid's ability to take up the nutrient iodide, which is a key building block for thyroid hormones. If the thyroid gland does not have enough iodide for a sufficient period of time, body's thyroid hormone levels will eventually drop. Hypothyroidism (low thyroid hormone levels) in adults can cause fatigue, depression, anxiety, unexplained weight gain, hair loss, and low libido. More serious, however, are the effects of thyroid hormone disruption in the developing fetus and child. Small changes in maternal thyroid hormone levels during pregnancy have been associated with reduced IQs in children [61, 62]. A recent epidemiological study by the US Centers for Disease Control (USCDC) shows that perchlorate exposures commonly found in the population can cause significant thyroid hormone disruptions in women – particularly in the population of women with lower iodine intake. Relying on a flawed industry study, the USEPA adopted a water clean-up standard for superfund sites of 24.5 ppb in 2006. Neither the USEPA nor the US Food and Drug Administration (USFDA) have taken any action to address the problem of widespread contamination in food.

Considering animal studies, perchlorate was first discovered to affect the thyroid in the 1950s, but it wasn't until the early 1990s that scientists began to conduct studies that involved feeding low doses of perchlorate to animals and looking for adverse effects. In 1995 the USEPA found that laboratory animals developed thyroid disorders after two weeks of drinking perchlorate-laced water. Subsequent studies found effects on brain and thyroid structure at even lower doses, and noted that rat pups born to exposed mothers were particularly like to show adverse effects [53, 54].

The USCDC conducted the first major epidemiological study on perchlorate exposure in the general population [59]. After testing urine samples of 2299 men and women from around the country for perchlorate, and comparing these findings with the levels of thyroid hormones found in the blood of these same people, the USCDC's researchers discovered that there was a statistically significant relationship between urinary perchlorate and thyroid hormone levels in the 1111 women tested. Furthermore, they found that if low iodine woman started with perchlorate exposure corresponding to 0.19 ppb in urine (the minimum level found), and then ingested enough perchlorate through food and/or drinking water to raise their urinary perchlorate level to 2.9 ppb (the median level found), their T4 thyroid hormone levels would drop by 13 percent. Similarly, if woman's urinary perchlorate level increased to 5.2 ppb (the 75th percentile exposure), their T4 levels would drop by 16 percent. These are significant declines when one considers that recent studies have shown that the cognitive development of the fetus is impaired in mothers with even mild disruptions in thyroid hormone levels [59, 61, 62]. Women with low iodine intake and levels of TSH (a type of thyroid hormone) that were already on the edge of the normal range were found to be even more sensitive to perchlorate exposure. For these women, if they were exposed to 5 parts per billion of perchlorate via food or drinking water, the resulting hormone disruption would push them into sub-clinical hypothyroidism.

4.1.5
Perfluorochemicals

The USEPA has described perfluorochemicals (PFCs) as combining "persistence, bioaccumulation, and toxicity properties to an extraordinary degree" [63]. PFCs are industrial chemicals widely used as water, stain, and grease repellants for food wrap, carpet, furniture, and clothing. The family includes such well known name brands as Scotchgard and Teflon.

PFCs are released to the environment in air and water emissions at numerous manufacturing and processing facilities worldwide. PFCs are also likely released to the environment at countless secondary manufacturing facilities, including sites where consumer products are coated for water, stain, and grease repellency. The dominant sources of PFCs in the environment are thought to be fluorotelomer chemicals, the active ingredients in coatings of furniture, clothing, food packaging, and other products. Fluorotelomers break down in the environment and in the body to PFCs differing only in the carbon chain length and end group [64, 65]. Most PFCs are fairly mobile in water, but due to low volatility of the persistent carboxy acids and sulfonates, many do not have the potential to migrate in air far from locations of release as a manufacturing pollutant. In contrast, studies indicate that PFC telomers are relatively volatile and could migrate long distances through the atmosphere.

Fluorotelomers are a likely source of the persistent perfluorochemicals found in newborns, and in wildlife and water in areas remote from manufacturing sites and human populations. Available scientific findings to date show that PFCs widely contaminate human blood [66, 67] and persist in the body for decades [68]. They act through a broad range of toxic mechanisms of action to present potential harm to a wide range of organs (ovaries, liver, kidney, spleen, thymus, thyroid, pituitary, testis), and persist indefinitely in the environment with no known biological or environmental breakdown mechanism [69–71]. Considering their ecotoxicity the newest evidence suggests that PFCc are able to induce and inhibit the activity of xenobiotic efflux transport proteins in marine bivalves [72].

4.1.6
Phthalates

Found within many consumer products, phthalates are industrial plasticizers that impart flexibility and resilience to plastic. They are common additives to soft plastic, especially PVC. They are present in clear food wrap, personal care products (detergents and soaps), and pesticides [73].

Phthalates are widely detected in human blood and urine samples. The latest exposure study from USCDC indicates that women are slightly more exposed than men, and younger children (ages 6–11) are more exposed than older children (ages 12–20) [74]. Exposure to phthalates occurs through direct use of cosmetics and other consumer products containing these chemicals, consumption of foods wrapped in products containing these chemicals, and through inhalation of air contaminated with these chemicals [74].

In laboratory animals, fetal exposure to phthalates causes significant developmental toxicity, especially of the male reproductive system. In adult animals, phthalates damage the reproductive organs, adrenal, liver, and kidney [75]. In utero exposure to high levels of phthalate metabolites are associated with marked differences in the reproductive systems of baby boys; the exposure levels associated with these health effects were not extreme, but rather were typical for about one-quarter of all women. Adult men with high levels of phthalates have lower sperm motility and concentration and alterations in hormone levels [76–78]. Concentrations of two phthalates in house dust are associated with asthma and rhinitis in a study of 400 children, half of whom had allergies [79].

4.1.7
Polybrominated Diphenyl Ethers

Polybrominated diphenyl ethers (PBDEs) are brominated fire retardants, intentionally added to flexible foam furniture, primarily mattresses, couches, padded chairs, pillows, carpet padding and vehicle upholstery, and to electronic products.

Studies of laboratory animals link PBDE exposure to an array of adverse health effects including thyroid hormone disruption, permanent learning and memory impairment, behavioral changes, hearing deficits, delayed puberty onset, decreased sperm count, and fetal malformations [80–82]. Research in animals shows that exposure to brominated fire retardants in utero or during infancy leads to more significant harm than exposure during adulthood, and at much lower levels [47]. PBDEs are bioaccumulative and lipophilic, and, therefore, are highly persistent in people and the environment. The chemicals build up in the body, are stored in fatty tissues and body fluids, such as blood and breast milk, and can be passed on to fetuses and infants during pregnancy and lactation. People are primarily exposed to PBDEs in their homes, offices, and vehicles. Secondary sources are foods, primarily meat, dairy, fish, and eggs [83].

Some PBDEs were withdrawn from the US market in 2005 due to their toxicity to laboratory animals, and their detection as contaminants in humans, wildlife, house and office buildings, and common foods [84–86]. Deca (PBDE-209), the form used in electronics, continues to be used in televisions, computer monitors and other electronic products. There is widespread concern that Deca breaks down in the environment to more toxic and persistent forms.

4.1.8 Polychlorinated Naphthalenes

There are 75 possible chemical variations of polychlorinated naphthalenes (PCNs). They have been used as cable insulation, wood preservatives, engine oil additives, electroplating masking compounds, capacitors, and in dye production. Products are generally mixtures of several different PCNs. The largest source of PCNs believed to be waste incineration and disposal of items containing PCNs, although other potential sources of PCNs to the environment include sewage discharge from municipal and industrial sites leaching from hazardous waste sites. PCNs are also unwanted byproducts formed after the chlorination of drinking water [87]. They have not been used commercially in significant quantities since the 1980s.

PCNs are toxic, persistent and bioaccumulate in people and wildlife. The toxic effects of many PCNs are thought to be similar to dioxin. In humans, severe skin reactions (chloracne) and liver disease have both been reported after occupational exposure to PCNs. Other symptoms found in workers include cirrhosis of the liver, irritation of the eyes, fatigue, headache, anaemia, haematuria, impotentia, anorexia, and nausea. At least ten deaths were reported from liver toxicity. Workers exposed to PCNs also have a slightly higher risk of all cancers combined [88–90].

4.2 Personal Care Products (PCPs)

4.2.1 Fragrances – Nitromusks and Polycyclic Musks

Nitromusk and polycyclic musks are synthetic fragrances typically used in cosmetics, perfume, air fresheners, cleansing agents, detergents, and soap. Musks are also used as food additives, in cigarettes, and in fish baits. Commonly used musks contaminate lakes and fish in the US and Europe [91–96]. Nitromusk and polycyclic musks tend to accumulate in the fatty tissues of our bodies, and are often detected in breast milk as well as blood [96–98].

In laboratory studies, some nitromusks have been linked to cancer [99, 100]. Studies of nitromusks in people suggest that high levels of some of these chemicals are associated with reproductive and fertility problems in women [101]. Some also produce skin irritation and sensitization [102, 103].

Growing concerns about the health effects of nitromusks have led the EU to ban the use of some of these chemicals in cosmetics and personal care products. As a result, the use of polycyclic musks has increased. However, laboratory studies suggest that polycyclic musks, like nitromusks, may also affect hormone systems [104–109]. Two particular musk chemicals, a nitromusk and a polycyclic musk which both produced neurotoxic effects in laboratory animals, have been removed from the market. In the US, all musk chemicals are unregulated, and safe levels of exposure have not yet been set. Considering their ecotoxic potential, Luckenbah and Epel [110] demonstrated that nitromusk and polycyclic musk compounds act as long-term inhibitors of cellular multixenobiotic resistance (MXR) defense systems mediated in aquatic mollusks by specific transport proteins.

4.2.2
Triclosan

Triclosan is essentially a pesticide (antibacterial agent), used in some healthcare facility soaps. It is also the most common antimicrobial agent in household liquid hand soap. It can be found in toothpaste, lip gloss, soap (solid and liquid), plastic products ranging from children's toys to cutting boards, and footwear [111]. It has been detected in human breast milk and serum samples from the general population [98, 112], and in the urine of 61% of 90 girls ages six to eight tested in a recent study spearheaded by Mount Sinai School of Medicine [73].

Triclosan kills microbes by disrupting protein production, binding to the active site of a critical carrier protein reductase essential for fatty acid synthesis, which is present in microbes but not humans. Available studies do not raise major concerns for human health, but some basic questions remain, including the safety of triclosan exposures in utero, and exposures in infancy through contaminated breast milk. Triclosan breaks down in the environment, including in tap water, to chlorinated chemicals that pose both environmental and health concerns [113].

Large quantities of triclosan are washed down drains and into wastewater treatment plants. A fraction is removed during water treatment, but the rest is discharged to lakes and rivers. Studies indicate that its interaction with sunlight results in the formation of methyl triclosan, a chemical that may bioacummulate in wildlife and humans [112, 114], as well as a form of dioxin, which is a chemical linked to a broad range of toxicities including cancer [115]. The Canadian government limits the levels of dioxins and furans allowed as impurities in personal care products that contain triclosan. Triclosan was recently found in 58% of 139 US streams [116], the likely result of its presence in treated discharged wastewater. A safety standard for triclosan has not yet been set, and it does not require testing in tap water. However, it is believed that triclosan likely passes through standard water treatment processes to contaminate treated tap water supplies at low levels. New studies show that triclosan in tap water will readily react with residual chlorine from standard water disinfecting procedures to form a variety of chlorinated byproducts, including chloroform, a suspected human carcinogen [117].

Wildlife species are also contaminated with triclosan and its breakdown products; a recent European study found its breakdown product methyl triclosan in fish, especially concentrated in fatty tissue [113]. Triclosan is known to be acutely toxic to certain types of aquatic organisms, but little is known about its long-term effects on humans [118]. The chemical structure of triclosan is similar to that of diethylstilbestrol (DES), a non-steroidal estrogen, raising concerns about its potential to act as an endocrine disruptor. A recent study showed that triclosan can affect the thyroid gland, significantly altering frog metamorphosis at exposure levels equivalent to those currently found in the environment and human tissues, suggesting that triclosan may represent a potential health risk to human hormone action as well [119]. Studies have also found that triclosan has weakly androgenic effects but no estrogenic effects [120]. In addition, animal studies have shown that prolonged application of triclosan solution to the skin can cause dermal irritation in people with a specific sensitivity. There is no evidence that triclosan is a carcinogen or teratogen [121]. There is concern that the widespread use of antimicrobials such as triclosan in household products may promote antibiotic resistance in bacteria, although the current literature shows a possible association but no definitive link [122].

In addition to the PCPs mentioned above, some other categories like sunscreen agents, preservatives, and nutraceuticals recently got attention as possible emerging contaminants. As for now, however, the weight of evidence does not justify their treatment as immediate hazard to human or wildlife health.

4.3 Pharmaceuticals (Human Drugs and Veterinary Medicines)

Recent studies have also identified a number of pharmaceuticals as potential environmental contaminants that may adversely affect reproduction and development of biota in the environment [111, 123]. Some of these substances are not removed in traditional, or even advanced treatment systems, or under best management practices [124, 125]. Several of these substances have recently been detected in well treated effluents and drinking water, showing that sewage treatment frequently does not affect the chemical structure, and, therefore, the toxicity of drugs [126–129]. Emerging data in Europe and North America suggests that these chemicals are widespread in the environment, especially in surface waters exposed to human or agriculture wastes [116, 130]. Consequently, pharmaceuticals often enter the environment at levels similar to better studied agrochemicals.

Traditionally, pharmaceuticals and personal care products have not been viewed as environmental pollutants [131]. However, the potential for these substances to cause a variety of physiological responses in non-target species has raised concerns for possible impacts on the environment. Although these substances are usually found at very low concentrations in the environment, continuous low-dose exposure to these complex mixtures, especially at sensitive life stages, may have significant effects on individuals, populations, or ecosystems. The ecological impact of long-term exposure to large mixtures of those essentially biologically active chemicals is also unknown. Many of these chemicals are known to be persistent in both treatment systems and in the environment. Chemicals found in sewage and manure, such as synthetic estrogens, are known to have biological consequences at extremely low exposures [132]. Exposure of biota to even low doses during critical or sensitive life-stages may have profound effects on development and reproduction for multiple generations.

Due to their intended use in human or veterinary medicine, pharmaceuticals are generally well studied and a large body of toxicological evidence directed to human health issues exists for most of them. Considering their ecotoxicity, however, the available evidence in most cases provides indications of acute effects in vivo for organisms at different trophic levels after short-term exposure, but extremely rarely after long-term chronic exposures. An excellent service called "The Pharmaceuticals in the Environment, Information for Assessing Risk" has been recently developed and is maintained at the National Centers for Coastal Ocean Science (NOAA), Center for Coastal Environmental Health and Biomolecular Research, USA [133]. The database provides information on prescribed amounts, levels detected in aquatic environments, chemical structure, molecular weight, octanol-water partition coefficients, water solubility, environmental persistence, general toxicity information, and specific toxicity levels of pharmaceuticals to five groups of organisms (algae, mollusks, finfish, crustaceans, and select terrestrial animals). Toxicity to terrestrial animals is provided as a general comparison to data found in toxicological literature. All of this information was obtained from available scientific literature and is provided to assist with indentification of locations where risks to aquatic organisms might occur.

Considering the ecotoxicity of human pharmaceuticals, most of the current knowledge is well summarized in several excellent review articles published during the last few years [111, 130, 134–136]. Summarizing the available data, it is clear that there is almost no data about bioaccumulation of pharmaceuticals in biota, and often there is no correlation between the acute toxicity and lipophilicity. Most of pharmaceuticals displayed their LC50 values above 100 mg/L, which classifies them as not being harmful to aquatic organisms. However, variability of data within the same, as well as between different species is considerable, often spanning one or two orders of magnitude. Nevertheless, the overall conclusion is that acute toxicity of pharmaceuticals may be only relevant in case of accidental spills. Chronic toxicity, however, appears to be more relevant to aquatic biota and numerous examples clearly point out that it cannot be derived from acute toxicity data by simple calculations.

Veterinary pharmaceuticals, on the other hand, were traditionally less covered in environmental and human health toxicity studies. Current livestock and aquaculture production practices include the use of a wide variety of pharmaceuticals to enhance animal health and efficient food production, including antimicrobials (antibiotics), growth enhancers, feed supplements, and other medicinal products. Recently, low levels of veterinary medicines were detected in soils, surface waters, and ground waters worldwide [137]. Although the environmental occurrence and associated impacts of some compounds, such as selected antibacterial compounds, have been investigated, the impacts of many other substances found in the environment are not well understood. As a result, questions have arisen about the effects of veterinary medicines on organisms in the environment and on human health.

The interest in veterinary pharmaceuticals as potential emerging contaminants has also stemmed from the proliferation of large-scale animal feeding operations (AFOs) during the last decade. The large number of animals produced creates a proportionately large volume of animal waste and associated emerging contaminants. In a reconnaissance study of liquid waste at swine AFOs in Iowa and North Carolina, US, multiple classes of antibiotics were detected ranging from ppb to ppm concentrations [138]. Compilation of data from liquid waste from swine operations between 1998 and 2002 found one or more antibiotics present in all of the samples. The data from these studies demonstrate that veterinary pharmaceuticals are excreted and frequently occur at detectable levels ranging from ppb to ppm concentrations in liquid and solid waste.

Research to document the presence of antibiotics in fish hatchery recently revealed the occurrence and persistence of antibiotics in medicated feed used in fish hatcheries [139]. It was discovered that ormetoprim and sulfadimethoxine persisted in water for longer periods of time than oxytetracycline in fish hatcheries. Oxytetracycline was detected more frequently in the samples of the intensive hatcheries than samples from the extensive hatcheries. Sulfadimethoxine concentrations were greater in the intensive hatcheries than the extensive hatcheries, but persisted up to 40 days after treatment in both types of fish hatcheries. In addition, antibiotics were detected in untreated hatchery raceways, suggesting that recirculating water within a hatchery can lead to unintentional low-level exposure of antibiotics to healthy fish.

4.4 Nonculturable Biological Pathogens as Emerging Contaminants

Among the viruses infecting humans, many different types are excreted in high concentrations in the feces of patients with gastroenteritis or hepatitis and in lower concentrations in the feces or urine of patients with other viral diseases. Moreover, viruses are also present in healthy individuals, and, thus, high viral loads are detected in urban sewage and are regarded as environmental contaminants [140]. Some viruses, such as humanpolyomaviruses and some adenovirus strains, infect humans during childhood, thereby establishing persistent infections. In the case of many frequent adenoviral respiratory infections, viral particles may continue to be excreted in feces for months or even years afterward. There is available information about some waterborne pathogens, but the improvement in molecular technology for detecting viruses present in water has focused attention on new groups of viruses that could be considered emergent viruses in diverse geographical areas. Technical advances are then most readily associated with the concept of emergent microorganisms, which are defined as newly identified microorganisms, those already existent but characterized by a rapidly increasing incidence and/or geographical ambit, and those for which transmission through food or water has only recently been discovered. Several studies have confirmed that infectious diseases related to water are not only a primordial cause of mortality and morbidity worldwide but also that both the spectrum and incidence of many diseases related to water are increasing. Human polyomaviruses, hepatitis E virus (HEV), and human adenoviruses (HAdV) are three groups of viruses, which are being detected more often in the environment [141]. Adenoviruses, for example, are important human pathogens that are responsible for both enteric illnesses and respiratory and eye infections. Recently, these viruses have been found to be prevalent in rivers, coastal waters, swimming pool waters, and drinking water supplies worldwide. USEPA listed adenovirus as one of nine microorganisms on the Contamination Candidate List for drinking water, because their survival characteristic during water treatment is not yet fully understood. Adenoviruses have been found to be significantly more stable than fecal indicator bacteria and other enteric viruses during UV treatment, and adenovirus infection may be caused by consumption of contaminated water or inhalation of aerosolized droplets during water recreation.

In addition, many species of bacteria pathogenic to humans, such as *Legionella*, are thought to have evolved in association with amoebal hosts. Several novel unculturable bacteria related to *Legionella* have also been found in amoebae, a few of which have been thought to be causes of nosocomial infections in humans [142]. A recent study done by Berk and colleagues in 2006 [143] revealed that it is over 16 times more likely to encounter infected amoebae in cooling towers than in natural environments. Several identified bacteria have novel rRNA sequences, and most strains were not culturable outside of amoebae. Such pathogens of amoebae may spread to the environment via aerosols from cooling towers. Therefore, studies of emerging infectious diseases should strongly consider cooling towers as a source of amoeba-associated pathogens.

Additional example is Campylobacter(s), which are emerging as one of the most significant causes of human infections worldwide, and the role that waterfowl and the aquatic environment have in the spread of disease is beginning to be elucidated [144]. On a world scale, Campylobacters are possibly the major cause of gastrointestinal infections. They are common commensals in the intestinal tract of many species of wild birds, including waterfowl. They are also widely distributed in aquatic environments where their origins may include waterfowl as well as sewage effluents and agricultural runoff. Campylobacters have marked seasonal trends and in temperate aquatic environments they peak during winter, whereas spring-summer is the peak period for human infection. Campylobacter species may survive, and remain potentially pathogenic, for long periods in aquatic environments. The utility of bacterial fecal indicators in predicting the presence of campylobacters in natural waters is questionable. Viable but nonculturable Campylobacter cells may occur, but whether they have any role in the generation of outbreaks of campylobacteriosis is unclear. The routine detection of *Campylobacter* spp. in avian feces and environmental waters largely relies on conventional culture methods, while the recognition of a particular species or strain is based on serotyping and increasingly on molecular methods.

4.5 Antibiotic Resistance Genes

Antibiotic resistance genes (ARGs) are another type of "biological" emerging environmental contaminants. Along with nanoparticles, they may be classical examples of indirect toxicants. The primary health concern in the case of ARGs is related to adverse outcomes of antibiotic's exposures resulting in selection for pathogen resistance or alteration of microbial community structures. The occurrence of ARGs was recently demonstrated in various environmental compartments including river sediments, irrigation ditches, dairy lagoons, and the effluents of wastewater recycling and drinking water treatment plants [145]. Some of ARGs were also present in treated drinking water and recycled wastewater, suggesting that these are potential pathways for the spread of ARGs to and from humans. On the basis of recent studies, there is a need for environmental scientists and engineers to help address the issue of the spread of ARGs in the environment.

4.6 Nanomaterials

I close this section with nanomaterials – the concerns of the future and "real" emerging contaminants. Engineered nanomaterials are commonly defined as materials designed and produced to have structural features with at least one dimension of 100 nanometers or less. Such materials typically possess nanostructure-dependent properties (e.g., chemical, mechanical, electrical, optical, magnetic, biological), which make them desirable for commercial or medical applications. However, these same properties potentially may lead to nanostructure-dependent biological activity that differs from and is not directly predicted by the bulk properties of the constituent chemicals and compounds.

The potential for human and ecological toxicity associated with nanomaterials and ultrafine particles is a growing area of investigation as more nanomaterials and products are developed and brought into commercial use. To date, few nanotoxicology studies have addressed the effects of nanomaterials in a variety of organisms and environments. However, the existing research raises some concerns about the safety of nanomaterials and has led to increased interest in studying the toxicity of nanomaterials for use in risk assessment and protection of human health and the environment. A new field of nanotoxicology has been developed to investigate the possibility of harmful effects due to exposure to nanomaterials [146]. Nanotoxicology also encompasses the proper characterization of nanomaterials used in toxicity studies. Characterization has been important in differentiating between naturally occurring forms of nanomaterials, nano-scale byproducts of natural or chemical processes, and manufactured (engineered) nanomaterials. Because of the wide differences in properties among nanomaterials, each of these types of nanoparticles can elicit its own unique biological or ecological responses. As a result, different types of nanomaterials must be categorized, characterized, and studied separately, although certain concepts of nanotoxicology, primarily based on the small size, likely apply to all nanomaterials.

As materials reach the nanoscale, they often no longer display the same reactivity as the bulk compound. For example, even a traditionally inert bulk compound, such as gold, may elicit a biological response when it is introduced as a nanomaterial [147]. The earliest studies investigating the toxicity of nanoparticles focused on atmospheric exposure of humans and environmentally relevant species to heterogeneous mixtures of environmentally produced ultrafine particulate matter (having a diameter < 100 nm). These studies examined pulmonary toxicity associated with particulate matter deposition in the respiratory tract of target organisms [148–151]. Epidemiological assessments of the effects of urban air pollution exposure focusing on particulate matter produced as a byproduct of combustion events, such as automobile exhaust and other sources of urban air pollution, showed a link in test populations between morbidity and mortality and the amount of particulate matter [152, 153].

Laboratory-based studies have investigated the effects of a large range of ultrafine materials through in vivo exposures using various animal models as well as cell-culture-based in vitro experiments. To date, animal studies routinely show an increase in pulmonary inflammation, oxidative stress, and distal organ involvement upon respiratory exposure to inhaled or implanted ultrafine particulate matter. Tissue and cell culture analyses have also supported the physiological response seen in whole animal models and yielded data pointing to an increased incidence of oxidative stress, inflammatory cytokine production, and apoptosis in response to exposure to ultrafine particles [154–157]. These studies have also yielded information on gene expression and cell signaling pathways that are activated in response to exposure to a variety of ultrafine particle species ranging from carbon-based combustion products to transition metals. Polytetrafluoroethylene fumes in indoor air pollution are nano-sized highly toxic particles [158]. They elicit a severe inflammatory response at low inhaled particle mass concentrations, suggestive of an oxidative injury.

In contrast to the heterogeneous ultrafine materials produced incidentally by combustion or friction, manufactured nanomaterials can be synthesized in highly homogenous forms of desired sizes and shapes (e.g., spheres, fibers, tubes, rings, planes). Limited research on manufactured nanomaterials has investigated the interrelationship between the size, shape, and dose of a material and its biological effects, and whether a unique toxicological profile may be observed for these different properties within biological models. Typically, the biological activity of particles increases as the particle size decreases. Smaller particles occupy less volume, resulting in a larger number of particles with a greater surface area per unit mass and increased potential for biological interaction [159]. Recent studies have begun to categorize the biological response elicited by various nanomaterials both in the ecosystem and in mammalian systems. Although most current research has focused on the effect of nanomaterials in mammalian systems, some recent studies have shown the potential of nanomaterials to elicit a phytotoxic response in the ecosystem. In the case of alumina nanoparticles, one of the US market leaders for nano-sized materials, 99.6% pure nanoparticles with an average particle size of 13 nm were shown to cause root growth inhibition in five plant species [160].

Charge properties and the ability of carbon nanoparticles to affect the integrity of the blood-brain barrier as well as exhibit chemical effects within the brain have also been studied. Nanoparticles can overcome this physical and electrostatic barrier to the brain. In addition, high concentrations of anionic nanoparticles and cationic nanoparticles are capable of disrupting the integrity of the blood-brain barrier. The brain uptake rates of anionic nanoparticles at lower concentrations were greater than those of neutral or cationic formulations at the same concentrations. This work suggests that neutral nanoparticles and low concentration anionic nanoparticles can serve as carrier molecules providing chemicals direct access to the brain and that cationic nanoparticles have an immediate toxic effect at the blood-brain barrier [161, 162].

Tests with uncoated, water soluble, colloidal C60 fullerenes have shown that redox-active lipophilic carbon nanoparticles are capable of producing oxidative damage in the brains of aquatic species [161]. The bactericidal potential of C60 fullerenes was also observed in these experiments. This property of fullerenes has possible ecological ramifications and is being explored as a potential source of new antimicrobial agents [163]. Oxidative stress as a common mechanism for cell damage induced by nanoparticles and ultrafine particles is well documented; fullerenes are model compounds for producing superoxide. A wide range of nanomaterial species have been shown to create reactive oxygen species both in vivo and in vitro. Species which have been shown to induce free radical damage include the C60 fullerenes, quantum dots, and carbon nanotubes. Nanoparticles of various sizes and chemical compositions are able to preferentially localize in mitochondria where they induce major structural damage and can contribute to oxidative stress [164].

Quantum dots (QDs) such as CdSe QDs have been introduced as new fluorophores for use in bioimaging. When conjugated with antibodies, they are used for immunostaining due to their bright, photostable fluorescence. To date, there is not sufficient analysis of the toxicity of quantum dots in the literature, but some current studies point to issues of concern when these nanomaterials are introduced into biological systems. Recently published research indicates that there is a range of concentrations where quantum dots used in bioimaging have the potential to decrease cell viability, or even cause cell death, thus suggesting that further toxicological evaluation is urgently needed [165, 166]. However, the research also highlights the need to further explore the long-term stability of the coatings used, both in vivo and exposed to environmental conditions.

5 Discussion

5.1 Regulatory Perspective and Public Concerns

In 2004, the environmental campaign group World Wide Fund (WWF) tested the blood of government ministers from 13 EU Member States for chemicals that can negatively affect human health and wildlife. WWF found on average 37 out of the 103 tested substances in the ministers' blood [167]. Further, it is clear that the EU citizens are concerned. In a recent survey, the impact of chemicals used in everyday products came fifth in a list of 15 environmental issues of concern. When asked about which issue they feel they lack information, citizens cited chemicals first [168]. Do they have reason(s) to be concerned? Undoubtedly, the answer is positive – the overview of the "chemical world", which is in this chapter concentrated only to today's man-made emerging contaminants, clearly suggests that there are real human and environmental health problems that have to be addressed. Considering the issue of chemical contamination, all critical parties – regulators, risk managers, industry sector, politicians, and, finally, scientists – do not offer answers and solutions needed for citizens to be less concerned.

Contamination of water supplies is an evolving problem and will remain an issue as long as technological change continues. Some of the contaminants now being targeted by researchers may come out with a clean slate, while others may require additional scrutiny. One of the hopes of today's researchers is that more sophisticated science will help speed the process of identifying and remedying the problems, before damage to either human health or the environment occurs. In any case, science and regulation must continue to evolve and change, as it has been the case in the past few years, to respond to new needs presented by chemicals and our increasing knowledge of them. At present, however, regulatory communities are placed in a reactive, rather than proactive, position with respect to identifying contaminants and addressing public concern. The current lists of environmental pollutants evolved from those established in 1970s and are mainly focused on conventional "priority pollutants" often referred to as "persistent organic pollutants" (POPs). As was elaborated, these chemicals represent only a tiny part of potential pollutants [1, 2] and biological systems may obviously suffer exposure to many more chemicals stressors, only a small number of which is regulated. Therefore, only a small proportion of potentially hazardous chemicals is toxicologically evaluated, and even smaller number of them is officially regulated.

This position is further emphasized in situations where federal funding is provided only on a short-term basis and only for specifically identified research needs, which by definition are reactionary calls to fill data gaps. Although this approach generates short-term products for stakeholders, it often leads to fragmentary, low profile science. In the long term, such goal-oriented approach to environmental funding does not allow for exploratory research that can be used to anticipate future environmental issues. Unfortunately, in the US, for example, there is no competitive funding scheme for the discovery of new contaminants. In addition, no cohesive plan exists to proactively screen and identify all contaminants of potential concern. On the other hand, both Canada and the EU are actively developing plans that will place them in positions from which they can anticipate future environmental issues. The Registration, Evaluation, and Authorization of Chemicals (REACH) regulation in the EU is a good example [169]. Entered into force in June 2007, it requires that manufacturers of substances and formulators register and provide prescribed (eco)toxicological data for all substances with a volume >1 metric ton per year. In contrast, the USEPA has taken a different tack by sponsoring a voluntary program called the High Production Volume (HPV) Challenge Program [170]. Since the program's inception in 1998, >2200 chemicals have been "adopted" by chemical manufacturers and importers. Unfortunately, this number is small in comparison with the number of chemicals included under REACH, and >200 HPV chemicals are still without the promise of toxicity testing.

5.2 (Eco)toxicological Constraints

As may be realized from this overview, (eco)toxicologists often seems to know too little too late, and are far too slow to respond to numerous chemicals that enter the market every day. Moreover, most of (eco)toxicological testing is done using traditional acute toxicity test protocols. As was reliably demonstrated with pharmaceuticals, acute toxicity cannot always serve as a reliable proxy for chronic toxicity effects encountered in real environmental situations. Certain substances may elicit adverse effects weeks, months or years after exposure. Carcinogenicity is a classical example – an ultimate adverse outcome difficult to characterize regarding causal connections. Consequently, chronic exposure assessments cannot be avoided and proper toxicological characterization will probably continue to be a time-consuming process.

The array of chemicals in use will likely continue to diversify and grow with changing use patterns in human populations and animal production facilities. Rapid developments in the pharmaceutical industry will also continue to quickly add to the vast number of chemicals already entering the environment. Due to the ever-increasing potency and specificity of pharmaceuticals, new substances may be of even greater concern for the environment. New approaches for testing and new ways of thinking about new materials are also necessary. The diverse routes of exposure, including inhalation, dermal uptake, ingestion, and injection, can present unique toxicological outcomes that vary with the physicochemical properties of the nanoparticles in question.

The likelihood of constantly introducing new chemicals to commerce pose inevitable doubts as to whether the chemical-by-chemical approach to toxicological testing and regulation of water pollutants will continue to be sustainable. In the past, studies have focused on the effects of single chemicals because chemicals are usually regulated singly. However, chemicals are always present as complex mixtures, thus some might say the regulation approach is naïve. Thus scientists are increasingly focusing on the toxicity of mixtures of chemicals, acknowledging that the toxicity expressed may be a result of additive or multiplicative effects, depending on interactions with other chemicals present in the environment. Furthermore, the issue becomes even more complex taking into account potential toxicity of numerous metabolites being generated from parent compounds.

An alternative approach, formalized as "toxicity apportionment" has been recently proposed [2]. The main principle of this approach would be to assign toxicity according to the total numbers of stressors present, without the need to know their identities in advance. The apportionment approach is especially valuable in accounting for all toxicants sharing the same mechanism of action. As was proposed, water monitoring programs based on that framework should utilize biomarkers and biotests designed around evolutionary biochemical features and mechanisms of action rather than individual chemical entities. This approach may indeed be the best way to simultaneously account for multitude of contaminants having the same mechanism of action, chemicals newly introduced to the market, and pollutants of the future. Looking from the cost-benefit side and trying to obtain relevant toxicological answers in a short time, an efficient screening protocol similar to that shown in Fig. 1, may be based on the extensive use of a series of small scale and in vitro biotests, used to rapidly and sensitively screen for the presence of contaminants of concern, including emerging contaminants, addressing both acute and chronic toxicity and utilizing test species on different levels of biological organizations. It can be used for testing of single chemicals and complex environmental samples. Such a battery of mechanism-based bioassays could be easily incorporated into monitoring efforts.

Nevertheless, whilst they are able to indicate the presence of certain groups of substances in well understood media based on a toxic response, caution is needed in broadening the application of in vitro tests to complex media such as effluents. In vitro tests that typically utilize genetically modified cells, yeast, or bacterial strains, demonstrate promising advantages such as speed, low cost and the ability to give an indication of specific toxicity that usually is not expressed in acute toxicity tests. However, they have to date only been used to a limited extent on effluents, making interpretation of test results difficult or in some cases impossible. Additional experience will be essential to improve the interpretation of test results and their relationship to actual environmental impacts. At present, even the best validated in vitro bioassays are only suitable as an initial screening step to prioritize effluents or effluent fractions for further study. In vivo tests with carefully selected indicator species are more appropriate to assess direct toxicity and should preferably be used for risk assessment purposes. Furthermore, bioassays can give both false negative and false positive results. False negative results may fail to highlight real health or environmental risks; false positives may imply health or environmental risks where, in fact, there are none. Due to the high sensitivity of these tests, false positives are likely when applied to complex mixtures like effluents.

Therefore, methods are now available that detect tiny quantities of chemicals which may potentially be hazardous. However, questions remain about which chemicals are responsible when positive results are obtained from drinking water, wastewater, freshwater and seawater, soil, mud, or any other sample. For effluents, it is a challenge that samples generally contain many compounds,

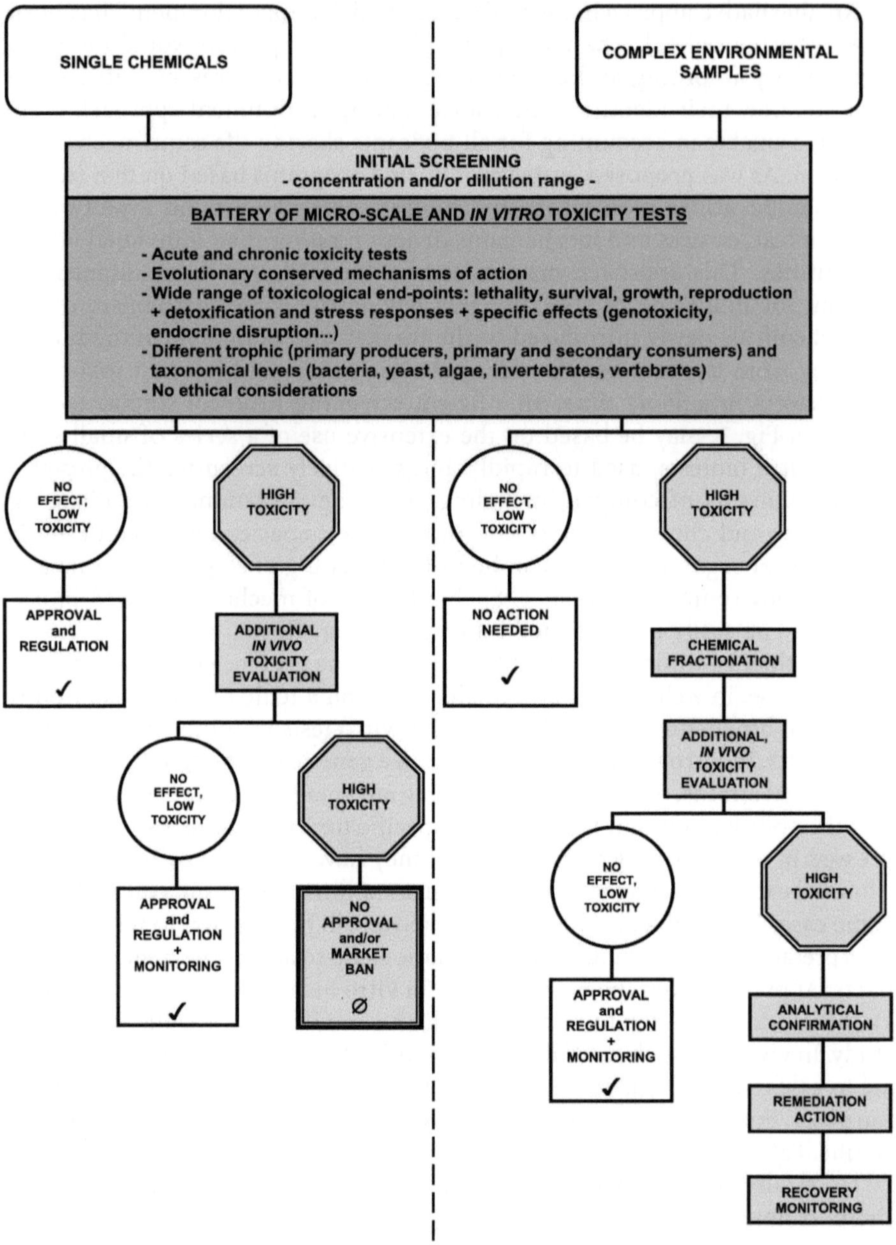

Fig. 1 Flow chart presentation of the possible (eco)toxicological protocol for rapid screening and characterization of single chemicals and complex environmental samples

resulting in false positives being frequently obtained. In the case of a positive response, the sample may be split up and analytical methods used to try and identify the responsive chemicals. Since these tests are highly sensitive and specific to the cell type used, the relevance of positive results for other species, living animals and longer-term exposures is the subject of ongoing studies. Consequently, a positive assay result should always be complemented with an in vivo assay and analytical detection to confirm the response. Only additional studies – coupled with a proper risk analysis, taking exposure into account – can confirm if the response indicates a genuine environmental risk.

Finally, regardless of the obstacles described, the most important concern regarding the exposure of aquatic and terrestrial organisms to emerging contaminants may be our inability to detect subtle health effects – imperceptible changes ranging from modification or reversal of attraction, behavioral changes related to feeding, matting, predator avoidance, or directional sensing. The changes we may see on the surface would simply be attributed to natural adaptation or any other form of natural changes. This concept of subtle changes, formalized at first by Kurelec in 1993 [171] as the genotoxic disease syndrome (GDS) was described as gradual accumulation of a wide spectrum of toxic events, none of which alone results in an easily detected adverse outcome. However, the final outcome would be an ultimate and often irreversible biological damage – species loss and decrease in biodiversity, unexpected and unexplained due to our inability to detect and act timely. These subtle, cumulative effects could make current toxicity-directed screening strategies largely useless in any effort to test waste effluents for toxic end points. At the moment, unfortunately, in the field of environmental toxicology there is no sound scientific answer to this critical issue. The raise of -omics techniques, however, especially genomics approach based on high-density microarray methodology, may be a future solution theoretically capable of detecting even subtle changes in gene expression patterns.

6 Conclusions and Future Directions

In this article, we briefly summarized major human end environmental health effects related to the most prominent categories of emerging contaminants, along with critical (eco)toxicological drawbacks and prerequisites needed for environmentally accountable risk characterization. The most important messages from this chapter, those we want for any reader to take into consideration are:

1. The threat posed by numerous emerging contaminants present in industrial and municipal waste is serious, poorly characterized, and should not be underestimated;

2. The research capacity of (eco)toxicology is at the moment far beyond capacity of analytical chemistry to detect new, emerging contaminants, and even more distant from the capacity of industry sector to design and introduce new chemical entities, likely "emerging" contaminants of the future;
3. Chronic, low-level exposure assessments do not have any scientifically sound alternative and should represent obligatory part of (eco)toxicity characterization of single chemicals and complex environmental mixtures;
4. The necessary improvements in the field of (eco)toxicology will not be possible without major shift in the regulatory arena, including significant changes in the environmental funding schemes.

Countries that adopt proactive approaches, such as the EU REACH initiative, will be afforded distinct environmental, economic, and scientific advantages, because they will be better serving human and nonhuman populations and ecosystems, with tangible savings to the healthcare and environment protection costs. Without the adoption of proactive plans to identify contaminants before they emerge, regulatory communities that remain in reactionary modes will be unable to fully serve the needs of the populations they represent.

Acknowledgements Financial support by the EU 6th Framework Specific Targeted Research Project: *Reduction of environmental and health risks, posed by Emerging Contaminants, through advanced treatment of municipal and industrial wastes* (*EMCO*; Contract No. INCO CT 2004-509188) is acknowledged. In addition, this work was partially supported by the Ministry of Science, Education and Sports of the Republic of Croatia, Project No: 098-0982934-2745 and 098-0982934-2712.

References

1. Daughton CG (2005) Renew Res J 21:6
2. Daughton CG (2004) Environ Imp Assess Rev 24:711
3. Boxall ABA, Sinclair CJ, Fenner K, Kolpin D, Maund SJ (2004) Environ Sci Technol 38:368A
4. The Human Toxome Project, Environmental Working Group, Washington DC, USA. Available at: http://www.ewg.org/sites/humantoxome/about/. Accessed 11 March, 2008
5. The Collaborative on Health and the Environment (CHE) Toxicant and Disease Database. Available et: http://database.healthandenvironment.org/. Accessed 11 March, 2008
6. USEPA (1980) Appendix B – Guidelines for Deriving Water Quality Criteria for the Protection of Aquatic Life and Its Uses. Federal Register, vol 45, No 231, November 28, 1980
7. Mount DI, Stephan CE (1967) Trans Am Fish Soc 96:185
8. McKim JM (1977) J Fish Res Board Can 34:1148
9. Macek KJ, Sleight BH (1977) Utility of toxicity tests with embryos and fry of fish in evaluating hazards associated with the chronic toxicity of chemicals to fishes. In: Mayer FL, Hamelink JL (eds) Aquatic Toxicology and Hazard Evaluation, ASTM STP 634. American Society for Testing and Materials, Philadelphia, p 137

10. Woltering DM (1984) Aquat Toxicol 5:1
11. USEPA (1981) In situ acute/chronic toxicological monitoring of industrial effluents for the NPDES biomonitoring program using fish and amphibian embryo/larval stages as test organisms. OWEP-82-00l. Office of Water Enforcement and Permits, US Environmental Protection Agency, Washington, DC 20460
12. Birge WJ, Black JA, Westerman AG (1985) Environ Toxicol Chem 4:807
13. Norberg TJ, Mount DI (1985) Environ Toxicol Chem 4:711
14. Birge WJ, Black JA, Ramey BA (1981) The reproductive toxicology of aquatic contaminants. In: Saxena J, Fisher F (eds) Hazard Assessments of Chemicals, Current Developments, vol 1. Academic Press, New York, p 59
15. Espejo R (2002) J Chromatogr A 976:335
16. Oros DR, Jarman WM, Lowe T, David N, Lowe S, Davis JA (2003) Mar Pollut Bull 46:1102
17. Saito I, Onuki A, Seto H (2004) Indoor Air 14:325
18. Laws SC, Carey SA, Ferrell JM, Bodman GJ, Cooper RL (2000) Toxicol Sci 54:154
19. Bechi N, Ietta F, Romagnoli R, Focardi S, Corsi I, Buffi C, Paulesu L (2006) Toxicol Sci 93:75
20. Kimura N, Kimura T, Suzuki M, Totsukawa K (2006) J Reprod Dev 52:789
21. McClusky LM, De Jager C, Bornman MS (2006) Toxicol Sci 95:249
22. Bistodeau TJ, Barber LB, Bartell SE, Cediel RA, Grove KJ, Klaustermeier J, Woodard JC, Lee KE, Schoenfuss HL (2006) Aquat Toxicol 79:268
23. Vom Saal F, Hughes C (2005) Environ Health Perspect 113:926
24. Howdeshell KL, Hotchkiss AK, Thayer KA, Vandenbergh JG, Vom Saal FS (1999) Nature 401:763
25. Sakaue LM, Ohsako S, Ishimura R, Kurosawa S, Kurohmaru M, Hayashi Y, Aoki Y (2001) J Occup Health 43:185
26. Al-Hiyasat AS, Darmani H, Elbetieha AM (2002) Eur J Oral Sci 110:163
27. Palanza PL, Howdeshell KL, Parmigiani S, Vom Saal FS (2002) Environ Health Perspect 110:415
28. Schonfelder G, Flick B, Mayr E, Talsness C, Paul M, Chahoud I (2002) Neoplasia 4:98
29. Wetherill YB, Petre CE, Monk KR, Puga A, Knudsen KE (2002) Mol Cancer Ther 1:515
30. Sugita-Konishi Y, Shimurab S, Nishikawab T, Sunagab F, Naitob H, Suzuki Y (2003) Toxicol Lett 136:217
31. Kabuto H, Amakawa M, Shishibori T (2004) Life Sci 74:2931
32. Della Seta D, Minder I, Dessì-Fulgheri F, Farabollini F (2005) Brain Res Bull 65:255
33. Markey CM, Wadia PR, Rubin BS, Sonnenschein C, Soto AM (2005) Biol Reprod 72:1344
34. Porrini S, Bellonia V, Della Seta D, Farabollini F, Giannelli G, Dessì-Fulgheri F (2005) Brain Res Bull 65:261
35. Timms BG, Howdeshell KL, Barton L, Bradley S, Richter CA, Vom Saal FS (2005) Proc Nat Acad Sci USA 102:7014
36. Alonso-Magdalena P, Morimoto S, Ripoll C, Fuentes E, Nadal A (2006) Environ Health Perspect 114:106
37. Olea N, Pulgar R, Pérez P, Olea-Serrano F, Rivas A, Novillo-Fertrell A, Pedraza V, Soto AM, Sonnenschein C (1996) Environ Health Perspect 104:298
38. Hanaoka T, Kawamura N, Hara K, Tsugane S (2002) Occup Environ Med 59:625
39. European Comission (2002) Study on the scientific evaluation of 12 substances in the context of endocrine disrupter priority list of actions – Final Report. WRc-NSF, UK. Available at: http://ec.europa.eu/environment/endocrine/documents/wrc_report.pdf#page=29

40. Inoue K, Yamaguchi A, Wada M, Yoshimura Y, Makino T, Nakazawa H (2001) J Chromatogr B 765:121
41. IARC (1999) Bisphenol A diglycidyl ether. IARC Monogr Eval Carcinog Risks Hum 71:1285
42. Warbrick EV, Dearman RJ, Ashbya J, Schmezer P, Kimber I (2001) Toxicology 163:63
43. Uter W, Rühl R, Pfahlberg A, Geier J, Schnuch A, Gefeller O (2004) Ann Occup Hyg 48:21
44. Ramilo G, Valverde I, Lago J, Vieites J, Cabado A (2006) Arch Toxicol 80:748
45. Satoh K, Ohyama K, Aoki N, Iida M, Nagai F (2004) Food Chem Toxicol 42:983
46. Viberg H, Fredriksson A, Jakobsson E, Örn U, Eriksson P (2003) Toxicol Sci 76:112
47. Viberg H, Johansson N, Fredriksson A, Eriksson J, Marsh G, Eriksson P (2006) Toxicol Sci 92:211
48. Eriksson P, Jakobsson E, Fredriksson A (2001) Environ Health Perspect 109:903
49. Viberg H, Eriksson P (2007) Neurotoxicology 28:136
50. Birnbaum LS, Staskal DF, Diliberto JJ (2003) Environ Int 29:855
51. Dasgupta PK, Dyke JV, Kirk AB, Jackson AW (2006) Environ Sci Technol 40:6608
52. USEPA (2005) Unregulated Contaminant Monitoring Program. US Environmental Protection Agency. Available at: http://www.epa.gov/safewater/ucmr/index.html. Accessed 11 March, 2008
53. EWG (2001) Rocket Science: Perchlorate and the toxic legacy of the cold war. Environmental Working Group, US. Available at: http://www.ewg.org/reports/rocketscience. Accessed 11 March, 2008
54. EWG (2003) Rocket Fuel in Drinking Water: New Studies Show Harm From Much Lower Doses. Environmental Working Group, US. Available at: http://www.ewg.org/node/8445. Accessed 11 March, 2008
55. EWG (2003). Suspect Salads: Toxic rocket fuel found in samples of winter lettuce. Environmental Working Group, US. Available at: http://www.ewg.org/reports/suspectsalads/ . Accessed 11 March, 2008
56. Kirk A, Smith EE, Tian K, Anderson TA, Dasgupta PK (2003) Environ Sci Technol 37:4979
57. Kirk A, Martinelango K, Tian K, Dutta A, Smith EE, Dasgupta PK (2005) Environ Sci Technol 39:2011
58. Sanchez CA, Crump KS, Krieger RI, Khandaker NR, Gibbs JP (2005) Environ Sci Technol 39:9391
59. Blount BC, Pirkle JL, Osterloh JD, Valentin-Blasini L, Caldwell KL (2006) Environ Health Perspect 114:1865
60. El Aribi H, Le Blanc YJC, Antonsen S, Sakuma T (2006) Anal Chim Acta 567:39
61. Haddow JE, Palomaki GE, Allan WC, Williams JR, Knight GJ, Gagnon J, O'Heir CE, Mitchell ML, Hermos RJ, Waisbren SE, Faix JD, Klein RZ (1999) N Engl J Med 341:549
62. Pop VJ, Kuijpens JL, Van Baar AL, Verkerk G, Van Son MM, De Vijlder JJ, Vulsma T, Wiersinga WM, Drexhage HA, Vader HL (1999) Clin Endocrinol 50:149
63. Auer C (2000) May 16, 2000 email message from Charles Auer (EPA) to OECD. EPA administrative record number AR226-0629
64. Hagen DF, Belisle J, Johnson JD, Venkateswarlu P (1981) Anal Biochem 118:336
65. Dinglasan MJ, Ye Y, Edwards EA, Mabury SA (2004) Environ Sci Technol 38:2857
66. Kannan K, Choi J-W, Isekic N, Senthilkumar K, Kima DH, Masunagac S, Giesy JP (2002) Chemosphere 49:225
67. Olsen GW, Burris JM, Lundberg JK, Hansen KJ, Mandel JH, Zobel LR (2002) Final Report: Identification of fluorochemicals in human sera. III. Pediatric participants in a group A streptococci clinical trial investigation US EPA Administrative Record

AR226-1085: Study conducted by Corporate Occupational Medicine. Medical Department, 3M Company, 220-3W-05, St Paul, MN, USA
68. Burris JM, Lundberg JK, Olsen GW, Simpson C, Mandel JH (2002) Determination of serum half-lives of several fluorochemicals. Interim Report No 2, St Paul, MN, 3M Company, US EPA docket AR-226-1086. US Environmental Protection Agency, Washington, DC
69. 3M (2000) Composite analytical laboratory report on the quantitative analysis of fluorochemicals in environmental samples. EPA Administrative Record AR226-0202, 3M
70. 3M (2001) Executive Summary: Environmental monitoring – multi-city study water, sludge, sediment, POTW effluent and landfill leachate samples
71. 3M (2001) Final Report, A longitudinal analysis of serum perfluorooctanesulfonate (PFOS) and perfluorooctanoate (PFOA) in relation to clinical chemistry, thyroid hormone, hematology and urinalysis results from male and female employee participants of the 2000
72. Stevenson CN, MacManus-Spencer LA, Luckenbach T, Luthy RG, Epel D (2006) Environ Sci Technol 40:5580
73. Wolff MS, Teitelbaum SL, Windham G, Pinney SM, Britton JA, Chelimo C, Godbold J, Biro F, Kushi LH, Pfeiffer CM, Calafat AM (2007) Environ Health Perspect 115:116
74. CDC (2005) National Report on Human Exposure to Environmental Chemicals. Centers for Disease Control, USA
75. CERHR (2000) NTP-CERHR expert panel report on di (2-ethylhexyl) phthalate (DEHP). Center for the Evaluation of Risks to Human Reproduction, USA
76. Duty SM, Barr DB, Brock JW, Ryan L, Chen Z, Herrick RF, Christiani DC, Hauser R (2003) Epidemiology 14:269
77. Duty SM, Calafat AM, Silva MJ, Brock JW, Ryan L, Chen Z, Overstreet J, Hauser R (2004) J Androl 25:293
78. Duty SM, Calafat AM, Manori SJ, Ryan L, Hauser R (2005) Hum Reprod 20:604
79. Bornehag C, Sundell J, Weschler CJ (2004) Environ Health Perspect 112:1393
80. Darnerud PO, Eriksen GS, Jóhannesson T, Larsen PB, Viluksela M (2001) Environ Health Perspect 109:49
81. Darnerud PO (2003) Environ Int 29:841
82. Hale RC, Alaee M, Manchester-Neesvig JB, Stapleton HM, Ikonomou MG (2003) Environ Int 29:771
83. Schecter A, Papke O, Tung KC, Joseph J, Harris TR, Dahlgren J (2005) J Occup Environ Med 47:199
84. Sjodin A, Patterson DG Jr, Bergman A (2001) Environ Sci Technol 35:3830
85. Sjodin A, Patterson DG Jr, Bergman A (2003) Environ Int 29:829
86. Sjodin A, McGahee EE III, Zhang Y, Turner WE, Slazyk B, Needham LL, Patterson DG Jr (2004) Environ Health Perspect 112:654
87. Vogelgesang J, Their HP (1986) Z Lebensm-Unters Forsch 182:400
88. Vinitskayaa H, Lachowicz A, Kilanowicz A, Bartkowiak J, Zylinska L (2005) Environ Toxicol Pharmacol 20:450
89. Van de Plassche EJ, Schwegler AM (2002) Polychlorinated naphthalenes. Dossier prepared for the third meeting of the UN-ECE Ad hoc Expert Group on POPs. Royal Haskoning report L0002.A0/R0010/EVDP/TL
90. Fromme H, Otto T, Pilz K (2001) Water Res 35:121
91. Peck AM, Hornbuckle KC (2004) Environ Sci Technol 38:367
92. Duedahl-Olesen L, Cederberga T, Høgsbro Pedersen K, Højgårdc A (2005) Chemosphere 61:422

93. Kannan K, Reiner JL, Yuna S, Perrotta EE, Tao L, Johnson-Restrepo B, Rodan BD (2005) Chemosphere 61:693
94. Peck AM, Linebaugh EK, Hornbuckle KC (2006) Environ Sci Technol 40:5629
95. Rimkus GG, Wolf M (1996) Chemosphere 33:2033
96. Liebl B, Mayer R, Ommer S, Sönnichsen C, Koletzko B (2000) Adv Exp Med Biol 478:289
97. Hutter HP, Wallner P, Moshammer H, Hartl W, Sattelberger R, Lorbeer G, Kundi M (2005) Chemosphere 59:487
98. TNO (2005) Man-made chemicals in maternal and cord blood. TNO Built Environment and Geosciences, Apeldoorn, The Netherlands, p 1
99. Maekawa A, Matsushima Y, Onodera H, Shibutani M, Ogasawara H, Kodama Y, Kurokawa Y, Hayashi Y (1990) Food Chem Toxicol 28:581
100. Apostolidis S, Chandra T, Demirhan I, Cinatl J, Doerr HW, Chandra A (2002) Anticancer Res 22:2657
101. Eisenhardt S, Runnebaum B, Bauer K, Gerhard I (2001) Environ Res 87:123
102. Parker RD, Buehler EV, Newmann EA (1986) Contact Dermatitis 14:103
103. Hayakawa R, Hirose O, Arima Y (1991) J Dermatol 18:420
104. Seinen W, Lemmen JG, Pieters RHH, Verbruggen EMJ, Van der Burg B (1999) Toxicol Lett 111:161
105. Chou YJ, Dietrich DR (1999) Toxicol Lett 111:27
106. Bitsch N, Dudas C, Körner W, Failing K, Biselli S, Rimkus G, Brunn H (2002) Arch Environ Contam Toxicol 43:257
107. Gomez E, Pillon A, Fenet H, Rosain D, Duchesne MJ, Nicolas JC, Balaguer P, Casellas C (2005) J Toxicol Environ Health A 68:239
108. Schreurs RH, Sonneveld E, Jansen JHJ, Seinen W, Van der Burg B (2005) Toxicol Sci 83:264
109. Schreurs RH, Sonneveld E, Van der Saag PT, Van der Burg B, Seinen W (2005) Toxicol Lett 156:261
110. Luckenbach T, Epel D (2005) Environ Health Perspect 113:17
111. Daughton CG, Ternes TA (1999) Environ Health Perspect 107:907
112. Adolfsson-Erici M, Parkkonen J, Sturve J (2002) Chemosphere 46:1485
113. Balmer ME, Poiger T, Droz C, Romanin K, Bergqvist P-A, Muller MD, Buser H-R (2004) Environ Sci Technol 38:390
114. Buser HR, Müller MD, Poiger T, Balmer ME (2002) Environ Sci Technol 36:221
115. Lores M, Llompart M, Sanchez-Prado L, Garcia-Jares C, Cela R (2005) Anal Bioanal Chem 381:1294
116. Kolpin D (2002) Environ Sci Technol 36:1202
117. Rule KL, Ebbet VR, Vikesland P (2005) Environ Sci Technol 39:3176
118. Orvos D, Versteeg VD, Inauen J, Capdevielle M, Rothenstein A, Cunningham V (2002) Environ Toxicol Chem 21:1338
119. Veldhoen N, Skirrow RC, Osachoff H, Wigmore H, Clapson DJ, Gunderson MP, Van Aggelen G, Helbing CC (2006) Aquat Toxicol 80:217
120. Foran CM, Bennett ER, Benson WH (2000) Marine Environ Res 50:153
121. Bhargava HN, Leonard PA (1996) Am J Infect Control 24:209
122. Russell AD (2002) Am J Infect Control 30:495
123. Ternes TA (1998) Water Res 32:3245
124. Ternes TA, Meisenheimer M, McDowell D, Sacher F, Brauch HJ, Haist-Gulde B, Preuss G, Wilme U, Zulei-Seibertet N (2002) Environ Sci Technol 36:3855
125. Webb S, Ternes T, Gibert M, Olejniczaket K (2003) Toxicol Lett 142:157
126. Ternes T, Kreckel P, Muelleret J (1999) Sci Total Environ 225:91

127. Ternes T, Stumpf M, Mueller J, Haberer K, Wilken RD, Servos M (1999) Sci Total Environ 225:81
128. Metcalfe CD, Koenig BG, Bennie DT, Servos M, Ternes TA, Hirschet R (2003) Environ Toxicol Chem 22:2872
129. Metcalfe CD, Miao XS, Koenig BG, Strugeret J (2003) Environ Toxicol Chem 22:2881
130. Halling-Sørensen B, Nors Nielsen S, Lanzky PF, Ingerslev F, Holten Lützhøft HC, Jørgensen SE (1998) Chemosphere 36:357
131. Hewitt LM, Servos MR (2001) Water Qual Res J Can 36:191
132. Metcalfe CD, Metcalfe TL, Kiparissis Y, Koenig BG, Khan C, Hughes RJ, Croley TR March RE, Potteret T (2001) Environ Toxicol Chem 20:297
133. Pharmaceuticals in the Environment, Information for Assessing Risk website. National Centers for Coastal Ocean Science, Center for Coastal Environmental Health and Biomolecular Research. Available et: http://www.chbr.noaa.gov/peiar/default.aspx. Accessed 11 March, 2008
134. Hirsch R, Ternes TA, Haberer K, Kratz KL (1999) Sci Total Environ 225:109
135. Damstra T, Barlow S, Bergman A, Kavlock R, Van der Kraak G (2002) Global assessment of the state of the science of endocrine disruptors. WHO/PCS/EDC/02.2
136. Fent K, Weston AA, Caminada D (2006) Aquat Toxicol 76:122
137. Boxall ABA, Kolpin DW, Halling-Sorensen B, Tolls J (2003) Environ Sci Technol 37:286A
138. Campagnolo ER, Johnson KR, Karpati A, Rubing CS, Kolpin DW, Meyer MT, Esteban JE, Currier RW, Smith K, Thu KM, McGeehin M (2002) Sci Total Environ 299:89
139. Thurman EM, Dietze JE, Scribner EA (2002) Occurrence of antibiotics in water from fish hatcheries. US Geological Survey Fact Sheet 120-02, p 4
140. Albinana-Gimenez N, Clemente-Casares P, Bofill-Mas S, Hundesa A, Ribas F, Girones R (2006) Environ Sci Technol 40:7416
141. Jiang SC (2006) Environ Sci Technol 40:7132
142. Hoge CW, Breiman RF (1991) Epidemiol Rev 13:329
143. Berk SG, Gunderson JH, Newsome AL, Farone AL, Hayes BJ, Redding KS, Uddin N, Williams EL, Johnson RA, Farsian M, Reid A, Skimmyhorn J, Farone MB (2006) Environ Sci Technol 40:7440
144. Abulreesh HH, Paget TA, Goulder R (2006) Environ Sci Technol 40:7122
145. Pruden A, Pei R, Storteboom H, Carlson KH (2006) Environ Sci Technol 40:7445
146. Donaldson K, Stone V, Tran CL, Kreyling W, Borm PJ (2004) Occup Environ Med 61:727
147. Goodman CM, McCusker CD, Yilmaz T, Rotello VM (2004) Bioconjugate Chem 15:897
148. Cheng YS, Hansen GK, Su YF, Yeh HC, Morgan KT (1990) Toxicol Appl Pharmacol 106:222
149. Bermudez E, Mangum JB, Wong BA, Asgharian B, Hext PM, Warheit DB, Everitt JI (2004) Toxicol Sci 77:347
150. Ferin J (1994) Toxicol Lett 72:121
151. Oberdorster G, Oberdorster E, Oberdorster J (2005) Environ Health Perspect 113:823
152. MacNee W, Donaldson K (2000) Monaldi Arch Chest Dis 55:135
153. Oberdorster G, Gelein RM, Ferin J, Weiss B (1995) Inhal Toxicol 7:111
154. Barlow PG, Donaldson K, MacCallum J, Clouter A, Stone V (2005) Toxicol Lett 155:397
155. Brown DM, Donaldson K, Borm PJ, Schins RP, Dehnhardt M, Gilmour P, Jimenez LA, Stone V (2004) Am J Physiol Lung Cell Mol Physiol 286:L344
156. Hetland RB, Cassee FR, Refsnes M, Schwarze PE, Lag M, Boere AJ, Dybing E (2004) Toxicol In Vitro 18:203

157. Stone V, Tuinman M, Vamvakopoulos JE, Shaw J, Brown D, Petterson S, Faux SP, Borm P, MacNee W, Michaelangeli F, Donaldson K (2000) Eur Respir J 15:297
158. De Hartog JJ, Hoek G, Peters A, Timonen KL, Ibald-Mulli A, Brunekreef B, Heinrich J, Tiittanen P, Van Wijnen JH, Kreyling W, Kulmala M, Pekkanen J (2003) Am J Epidemiol 157:613
159. Oberdorster G (1996) Inhal Toxicol 8:73
160. Warheit DB (2004) Mater Today 7:32
161. Oberdorster E (2004) Environ Health Perspect 112:1058
162. Lockman PR, Koziara JM, Mumper RJ, Allen DD (2004) J Drug Target 12:635
163. Yamakoshi YN, Yagami T, Sueyoshi S, Miyata N (1996) J Org Chem 61:7236
164. Li N, Sioutas C, Cho A, Schmitz D, Misra C, Sempf J, Wang M, Oberley T, Froines J, Nel A (2003) Environ Health Perspect 111:455
165. Lovric J, Bazzi HS, Cuie Y, Fortin GR, Winnik FM, Maysinger D (2005) J Mol Med 83:377
166. Shiohara A, Hoshino A, Hanaki K, Suzuki K, Yamamoto K (2004) Microbiol Immunol 48:669
167. WWF (2004) Bad blood? A Survey of chemicals in the blood of European ministers (http://www.worldwildlife.org/toxics/pubs/badblood.pdf). Accessed 11 March, 2008
168. The attitudes of European citizens toward the environment, Special Eurobarometer 217/Wave 62.1, conducted in November 2004, published in April 2005 (http://europa.eu.int/comm/environment/barometer/index.htm).
169. Registration, Evaluation, Authorisation and Restriction of Chemicals (REACH) system in the EU. Available at: http://ecb.jrc.it/reach/
170. High Production Volume (HPV) Challenge Program of the USEPA. Available at: http://www.epa.gov/hpv/. Accessed 11 March, 2008
171. Kurelec B (1993) Mar Environ Res 35:341
172. Chemical Abstracts Service (CAS) of The American Chemical Society. Available at: http://www.cas.org/index.html. Accessed 11 March, 2008

Hdb Env Chem Vol. 5, Part S/1 (2008): 143–168
DOI 10.1007/698_5_101

Published online: 24 January 2008

Traceability of Emerging Contaminants from Wastewater to Drinking Water

M. Huerta-Fontela[1,2] (✉) · F. Ventura[1]

[1]AGBAR-Aigües de Barcelona, Av. Diagonal 211, 08018 Barcelona, Spain
mhuerta@agbar.es

[2]Department of Analytical Chemistry, University of Barcelona, Av. Diagonal 647, 08028 Barcelona, Spain

Abstract Due to the incomplete elimination of some human contaminants during wastewater treatment, some of these compounds can be found in surface waters or groundwaters which are used as raw waters for drinking water production. The treatment efficiency to completely eliminate these emerging contaminants or to partially remove them will determine the quality of the final treated water. Up to today, few studies have been performed to evaluate the efficiency of the usual drinking water treatments in eliminating emerging contaminants. Moreover, every day new potential emerging contaminants are discovered and new disinfection by-products are also generated during treatment, with a total ignorance of their potential toxicity or effect on human health. In this chapter, a summary of the state of the art of emerging contaminant occurrence and elimination during drinking water processes at the bench scale or real scale is presented. A study of the presence and elimination of a new group of human contaminants, susceptible to being considered as a new emerging contaminant group, in a real drinking water treatment plant in Spain has also been included.

Keywords Carbon · Disinfection by-products · Drinking water · Emerging contaminants · Illicit drugs · Oxidation · Sorption

1 Emerging Contaminants in Drinking Water

The occurrence of emerging contaminants (i.e., human and veterinary drugs, surfactants, textile dyes, algal toxins, etc.) in wastewaters [1–7] and surface waters [1, 8–14] and their removal during conventional treatments has been widely evaluated in recent years. Several organic pollutants, e.g., pharmaceuticals, are not quantitatively eliminated by wastewater treatment and "survive" natural attenuation processes in surface waters. Therefore, the occurrence of these contaminants in these resources can have a negative impact on the quality of drinking water and, perhaps, produce adverse health effects. The incidence of these organic micropollutants in raw water and their elimination during drinking water treatment, as well as the formation of disinfection by-products (DBPs), are issues related to the quality of raw resources and water supplies. Compared to wastewater treatment plants, much less is known about the behavior of these compounds in drinking water treatment plants (DWTPs). In Table 1, a summary of some of the emerging contaminants detected in drinking water is displayed. The lack of systematic monitoring programs or the fact that they are present at fluctuating concentrations near the analytical method detection limits (some of these compounds usually occur in the low ng/L range) could be some reasons to explain the relatively little knowledge of the occurrence of these compounds in drinking water production [15]. However, several studies have found that the removal of emerging contaminants (mostly polar compounds) during drinking water treatment is incomplete. In 1993, clofibric acid, the active metabolite of some blood lipid regulators such as clofibrate, etofyllin clofibrate, and etofibrate, was found in Berlin tap water at high concentrations above 165 ng/L. Further studies, showed a direct correlation between bank filtration and artificial groundwater enrichment (used by a particular waterworks in drinking water production) and the concentrations of this drug in treated water [16, 17]. The same authors also detected the presence of propylphenazone and diclofenac in finished drinking water. Clofibric acid occurrence was also investigated in drinking waters from southern California [18]. This compound was not found in the samples analyzed; however, ibuprofen, triclosan, several phthalates, and additives were detected in samples of finished drinking water. These authors also performed a seasonal study to evaluate the performance of these compounds through time, concluding that higher concentrations in raw waters were detected between August and November (dry season), probably related to lower flow rates.

Boyd et al. [19] examined the occurrence of nine pharmaceuticals and personal care products (PPCPs) and endocrine disrupting compounds (EDCs), including clofibric acid, anti-inflammatories, analgesics, antibiotics, and hormones, in drinking water from the USA and Canada, and none of them was found in the finished drinking water. The presence of several pharmaceuticals, including lipid regulators, analgesics, anti-inflammatories, and their metabolites, was also evaluated in tap water from Cologne (Germany) [20]. Most of these compounds were found in the rivers and ponds analyzed but none of the eight selected drinking water samples showed the presence of the studied pharmaceuticals. Nevertheless, some hormones and antibiotics were detected in final drinking waters from the USA and Italy in recent years [21, 22]. Thus, McLachlan et al. [21] showed the presence of 17β-estradiol, estriol, and nonylphenol in final drinking waters. Regarding antibiotics, Perret et al. [22] studied the occurrence of 11 sulphonamide compounds (SAs) in mineral and municipal drinking waters from Italy. Concentrations of SAs from 9 to 80 ng/L in four different brands of mineral waters were obtained, while drinking water treatment was shown to be effective in the elimination of these compounds, with concentrations of SAs in municipal waters below the limit of quantification.

MTBE, a gasoline additive used since 1979, has also been detected in finished drinking water from the USA and Europe. Williams [23] reported the occurrence of this contaminant in about 1.3% of the drinking water samples from California (USA) analyzed during a period of 6 years. Concentrations ranged from 5 to 15 μg/L, nevertheless only 27% of the positive samples exceeded California's primary health-based standard of 13 μg/L. MTBE was also found in tap water from Germany; Achten et al. [24] reported maximum concentrations above 71 ng/L in treated water from the Frankfurt area. In 1997, another emerging contaminant, perchlorate, was discovered in water supplies from the USA. Exhaustive surveys were performed in California (USA) and perchlorate was found in 185 out of 2200 drinking water sources analyzed [25].

Algal toxins can also impact humans through drinking water contamination. The most lethal outbreak attributed to the presence of cyanobacteria in drinking water occurred in Brazil, where 88 deaths occurred over a 42-day period [26]. In 1999, toxic cyanobacteria blooms, microcystins, anatoxin-a, and cylindrospermopsin were also found in finished drinking waters from Florida (USA) at levels higher than those proposed in human health guidelines [27].

A more extended study was performed by Stackelberg et al. [28] who evaluated the persistence of 106 organic wastewater-related contaminants through conventional treatment processes and their occurrence in finished treated water. Results showed the presence of 17 of the selected contaminants in final water samples; caffeine (0.119 μg/L), carbamazepine (0.258 μg/L), dehydronifedipine (nifedipine metabolite; 0.004 μg/L), and cotinine (nicotine metabolite; 0.025 μg/L) were detected among the selected prescription and nonprescription drugs. Fragrances such as 7-acetyl-1,1,3,4,4,6-hexamethyl tetrahydronaphthalene (AHTN or Tonalide) and 1,3,4,6,7,8-hexahydro-4,6,6,7,8,8-hexamethylcyclopenta-γ-2-benzopyran (HHCB or Galaxolide), the cosmetic triethyl citrate, and the plasticizer bisphenol A were found at high ng/L concentrations. Some pesticides, flame retardants, and solvents were also detected

Table 1 Summary of emerging contaminants found in tap water reported in the literature

Compound	Classification	Concentration (in tap water)	Country	Refs.
Bezafibrate	Pharmaceutical	up to 27 ng/L	Germany	[4]
		0.7 ng/L	Italy	[29]
Carbamazepine	Pharmaceutical	up to 30 ng/L	Germany	[4]
		119 ng/L	USA	[28]
		43.2 ng/L	France	[30]
		up to 20 ng/L	Germany	[123]
		5 ng/L	Italy	[29]
		140 ng/L	USA	[31]
Clofibrate	Pharmaceutical	270 ng/L	Germany	[17]
		0.58 ng/L	USA	[18]
Clofibric acid	Pharmaceutical	70–7300 ng/L	Germany	[17]
		3.2–5.3 ng/L	Italy	[32]
		up to 70 ng/L	Germany	[4]
		0.63 ng/L	USA	[18]
Codeine		30 ng/L	USA	[31]
Diazepam	Pharmaceutical	19.6–23.5 ng/L	Italy	[32]
Diclofenac	Pharmaceutical	up to 6 ng/L	Germany	[4]
		0.4–0.9 μg/L	Germany	[33]
Dilantin	Pharmaceutical	1.3 ng/L	USA	[34]
Fenofibric acid	Pharmaceutical	up to 45 ng/L	Germany	[17]
		up to 42 ng/L	Germany	[4]
Gemfibrozil	Pharmaceutical	0.4 ng/L	Italy	[29]
		70 ng/L	Canada	[35]
Ibuprofen	Pharmaceutical	up to 200 ng/L	Germany	[17]
		up to 3 ng/L	Germany	[4]
		0.6 ng/L	France	[30]
		5.85 ng/L	USA	[18]
Ibuprofen methyl ester	Metabolite	9.22 ng/L	USA	[18]
Ketoprofen	Pharmaceutical	3.0 ng/L	France	[30]
Meprobamate	Pharmaceutical	5.9 ng/L	USA	[34]
Naproxen	Pharmaceutical	0.15 ng/L	France	[30]
Paracetamol	Pharmaceutical	211 ng/L	France	[30]
Phenazone	Pharmaceutical	up to 1250 ng/L	Germany	[17]
		up to 50 ng/L	Germany	[4]
		400 ng/L	Germany	[36]
		250 ng/L	Germany	[37]
Primidone	Pharmaceutical	up to 20 ng/L	Germany	[123]
Propiphenazone	Pharmaceutical	up to 1465 ng/L	Germany	[17]
		120 ng/L	Germany	[36]
		80 ng/L	Germany	[37]
Sulfamethizole	Pharmaceutical (veterinary)	9 ng/L	Italy	[22]
Sulfamethoxazole	Pharmaceutical (veterinary)	8–13 ng/L	Italy	[22]

Table 1 (continued)

Compound	Classification	Concentration (in tap water)	Country	Refs.
Sulfadimethoxine	Pharmaceutical (veterinary)	11 ng/L	Italy	[22]
Tylosin	Pharmaceutical	0.6–1.7 ng/L	Italy	[32]
Diatrizoic acid	X-ray contrast	up to 85 ng/L	Germany	[4]
Iopamidol	X-ray contrast	up to 79 ng/L	Germany	[4]
Iopromid	X-ray contrast	up to 86 ng/L	Germany	[4]
Caffeine	Stimulant	0.119 μg/L	USA	[28]
		0.237 μg/L	Italy	[29]
		0.06 μg/L	USA	[31]
Cotinine	Nicotine metabolite	25 ng/L	USA	[28]
		20 ng/L	USA	[31]
17α-Ethynilestradiol	Hormone	50 pg/L	Germany	[38]
Benzophenone	Sunscreen	0.13 μg/L	USA	[28]
Hydrocinnamic acid	Sunscreen	12.5 ng/L	USA	[18]
Triclosan	Germicide	0.734 μg/L	USA	[18]
		0.14 μg/L	USA	[8]
AHTN	Fragrance	0.49 μg/L	USA	[28]
		0.068 μg/L	USA	[31]
Camphor	Fragrance	0.017 μg/L	USA	[31]
HHCB	Fragrance	0.082 μg/L	USA	[28]
Triethyl citrate	Cosmetic	0.062 μg/L	USA	[28]
		0.082 μg/L	USA	[31]
MTBE	Gasoline additive	<13 μg/L	USA	[23]
		up to 75 ng/L	Germany	[24]
Anatoxin-A	Algal toxin	8.5 μg/L	USA	[27]
Cylindrospermopsin	Algal toxin	97.1 μg/L	USA	[27]
Microcystin	Algal toxin	up to 12.5 μg/L	USA	[27]
		up to 1 μg/L	USA	[39]
		<1 μg/L	Germany and Switzerland	[40]
Dimethyl phthalate	Plasticizer	2.36 μg/L	USA	[18]
Diethyl phthalate	Plasticizer	0.16–0.2 μg/L	Germany and Poland	[41]
		0.3 μg/L	Greece	[42]
		2.10 μg/L	USA	[18]
Dibutyl phthalate	Plasticizer	0.38–0.64 μg/L	Germany and Poland	[41]
		0.2–10.4 μg/L	Germany	[43]
		1.04 μg/L	Greece	[42]
		3.71 μg/L	USA	[18]
Butyl benzyl phthalate	Plasticizer	0.02–0.05 μg/L	Germany and Poland	[41]
		0.7 μg/L	Germany	[43]
		0.651 μg/L	USA	[18]
DEHP	Plasticizer	0.05–0.06 μg/L	Germany and Poland	[41]
		0.93 μg/L	Greece	[42]

but their concentrations did not exceed the maximum concentration levels established by the US Environmental Protection Agency.

Recently, Loos et al. [29] performed a survey of the anthropogenic environmental pollutants in surface and drinking waters from Italy. Fifty-one contaminants including pharmaceuticals, hormones, phthalates, surfactants, and herbicides were analyzed in both water matrices. Results achieved from surface waters coming from a lake showed the presence of 28 contaminants in the ng/L concentration range and similar concentration levels were obtained in tap water for the 23 detected compounds. For instance, pharmaceuticals such as carbamazepine, gemfibrozil, and bezafibrate were found at 5, 0.4, and 0.7 ng/L concentration levels in the tap water samples analyzed.

2 Emerging Contaminants During Drinking Water Treatment

2.1 Activated Carbon Adsorption

Activated carbon is a commonly used adsorbent for the removal of organic compounds such as pesticides, pharmaceuticals, and odor and taste compounds [44–46]. Adsorption on activated carbon depends on the intrinsic properties of the activated carbon sorbent (surface area and charge, pore size distribution, oxygen content) and on the solute properties (shape, size, charge, and hydrophobicity). Removal of such organic compounds is mainly controlled by hydrophobic interactions.

Powdered activated carbon (PAC) was evaluated for the elimination of selected PPCPs and endocrine disruptors during simulated drinking water treatment processes in the laboratory [47]. Octanol–water partition coefficients were shown to be a reasonable indicator of compound removal in PAC test conditions. Therefore, compounds with $\log K_{\mathrm{OW}}$ values higher than 3 (i.e., sulfamethoxazole or carbamazepine) showed elimination percentages higher than 70% (5 mg/L; 4-h contact time) except for compounds with deprotonated acid functional groups (i.e., naproxen or ibuprofen), which seemed the most difficult to remove with PAC. Deviations from this correlation were also detected for N-heterocyclic compounds (i.e., caffeine or trimethoprim) or protonated bases (i.e., acetaminophen) with low K_{OW}, which showed higher removal percentages than expected.

Granular activated carbon (GAC) was also evaluated for the elimination of pharmaceuticals (bezafibrate, clofibric acid, diclofenac, and carbamazepine) under laboratory, pilot, and waterworks conditions in Germany [33]. Pilot experiments showed high adsorption capacities for all the compounds except for clofibric acid, which due to its acidic properties had a low breakthrough volume (17 m^3/kg in a 160-cm carbon layer). In waterworks, GAC filtration was also shown to be a very effective removal process, even at high concentrations of pharmaceuticals. They were almost completely removed at throughputs over 70 $m^3 kg^{-1}$ except for clofibric acid, which could be removed completely at 15–20 $m^3 kg^{-1}$. Nevertheless, the results obtained for carbamazepine were contradicted by a subsequent study performed in a DWTP in the USA [28]. In this work, GAC efficiency was evaluated for the elimination of prescription and nonprescription drugs, fragrance compounds, PPCPs, and other organic contaminants. These studies indicated that carbamazepine and other hydrophobic compounds, such as fragrances HHCB (Galaxolide) and AHTN (Tonalide), persisted through DWTPs including filtration with GAC. The authors suggested that different sorption efficiencies depend on competition with other organic compounds; therefore, the adsorption capacities for these compounds result in smaller values in a DWTP that contains amounts of organic compounds rather than in a laboratory or pilot-scale experiment.

2.2 Oxidation Processes

In drinking water treatment systems, the oxidants commonly used are chlorine, chlorine dioxide, and ozone. Ozone is widely used in Europe for the treatment of surface waters while free chlorine is preferred in the USA, although in recent years ozone use has experienced an increase. All three oxidants are strong electrophiles that exhibit selective reactivity with organic compounds. Among them, ozone tends to be more reactive, following the order $O_3 > ClO_2 > HOCl$. One exception is waters with high ammonia content where chlorine has the highest reactivity.

Oxidation processes have to deal with one major drawback, the formation of undesirable DBPs which in some cases can exhibit higher toxicity than the precursors. A summary of some DBPs from pharmaceuticals produced during oxidation processes described in the literature is displayed in Table 2.

2.2.1 Ozonation

Ozone is used in water treatment as both disinfectant and oxidant and reacts with a large number of organic and inorganic compounds [48–50]. Rate constants for the reaction with ozone range several orders of magnitude, showing that ozone is a very selective oxidant. Regarding organic compounds, ozone is particularly reactive toward amines, phenols, and compounds with double bonds, especially in aliphatic compounds. In addition, ozone is unstable in water (from seconds to hours) and its decomposition leads to a major secondary oxidant, the hydroxyl radical [50, 51]. The OH radical is a powerful but less selective oxidant; it reacts with high rate constants with most organic compounds but these reactions are less efficient because a large fraction is scavenged by

Table 2 Summary of DBPs described in the literature and produced from pharmaceutical precursors during oxidation processes

Compound	Class	Process	DBPs/Intermediates	Refs.
Amoxicillin	Antibiotic	Ozonation	Elemental sulfur	[83]
		Ozonation	Hydroxylation of phenol ring	[84]
MMTD	Antibiotic	H_2O_2/UV	Degradation pathway	[85]
		Photolysis	One DBP	[85]
MMTD-Me	Antibiotic	H_2O_2/UV	S oxidation	[86]
		Photolysis	Two DBPs	[86]
Lincomycin	Antibiotic	TiO_2/hν	Mineralization	[87]
Sulfadiazine	Antibiotic	TiO_2/hν	4-Methyl-2-aminopyrimidine	[88]
Sulfamethoxazole	Antibiotic	Ozonation	Hydroxylamine formation	[60]
		Chlorination	3-Amino-5methylisoxazole; *N*-chloro-*p*-benzoquinoneimine	[75]
Sulfamethoxine	Antibiotic	TiO_2/hν	2,6-Dimethoxy-4-aminopyrimidine; 2-aminothiazole	[88]
Sulfamerazine	Antibiotic	TiO_2/hν	4-Methyl-2-aminopyrimidine	[88]
Sulfathiazole	Antibiotic	TiO_2/hν	2,6-Dimethoxy-4-aminopyrimidine; 2-aminothiazole	[88]
Trimethoprim	Antibiotic	Ozonation/chlorination	Degradation pathway	[89]
		Chlorination	Chlorinated and hydroxylated products from TMP's 3,4,5-trimethoxybenzyl moiety	[90]
Busperidone	Antianxiety	TiO_2/hν	Hydroxybusperidone; dihydroxybusperidone, dipyrimidinylbusperidone; 1-pyrimidinyl piperazine	[91]
Carbamazepine	Anticonvulsant	Ozonation	Degradation pathway	[92]
		Ozonation	1-(2-Benzaldehyde)-4-hydro-(1*H*,3*H*)-quinazoline-2-one (BQM) 1-(2-Benzaldehyde)-(1*H*,3*H*)-quinazoline-2,4-dione (BQD) 1-(2-Benzoic acid)-(1*H*,3*H*)-quinazoline-2,4-dione (BaQD)	[57]
		H_2O_2/UV	Acridine, salicylic acid, catechol, and anthranilic acid	[93]
		TiO_2/hν	Degradation pathway	[94]

Table 2 (continued)

Compound	Class	Process	DBPs/Intermediates	Refs.
Acetaminophen	Analgesic	Chlorination	Chloro-4-acetamidophenol; dichloro-4-acetaminophenol, 1,4-benzoquinone; *N*-acetyl-*p*-benzoquinone imine	[72]
Diclofenac	Anti-inflammatory	Ozonation	Degradation pathway	[95]
		H_2O_2/UV	Degradation pathway	[95]
		Photo-Fenton	Degradation pathway	[96]
Paracetamol	Anti-inflammatory	Ozonation	*N*-(4-hydroxyphenyl) acetamide	[92]
		Ozonation	*N*-(4-hydroxyphenyl) acetamide	[57]
		H_2O_2/UV	2-[(2,6-Dichlorophenyl)amino]-5-hydroxyphenylacetic acid); 2,5-dihydroxyphenylacetic acid	[97, 98]
		Anodic oxidation	Oxalic acid, oxamic acid	[99]
Cimetidine	Histamine receptor	Fenton	Cimetidine sulfoxide, *N*-desmethylcimetidine, *N*-desmethylcimetidine sulfoxide, cimetidine guanylurea, and 5-hydroxymethylimidazole	[100]
Ranitidine	Histamine receptor	$TiO_2/h\nu$	Mineralization	[101]
17β-Estradiol	Hormone	Ozonation	Degradation pathway	[71]
		Ozonation	Oxidized and chlorine substituted compounds	[72]
		$TiO_2/h\nu$	10ε-17β-Dihydroxy-1,4-estradien-3-one and testosterone-like species	[102]
Estrone	Hormone	Ozonation	Degradation pathway	[70]
		Photo-Fenton	Six intermediates	[102]
		Chlorination	2-Chloroestrone, 4-chloroestrone, 2,4-dichloroestrone, and 1,4-estradiene-3,17-dione	[103]
17α-Ethinylestradiol	Hormone	Ozonation	Dehydrated and decarboxylated compounds (five products)	[71]
		Ozonation	Oxidized and chlorine substituted compounds	[72]
Clofibric acid	Lipid regulator	$TiO_2/h\nu$	Degradation pathway	[94]
Iomeprol	X-ray contrast	$TiO_2/h\nu$	By-products unidentified	[94]

the water matrix. Additional oxidation processes are the advanced oxidation processes (AOPs) which use OH radicals as the main oxidants. These processes accelerate the formation of radicals by increasing the pH in water, by adding hydrogen peroxide, or by applying UV radiation [50, 52, 53].

Several experiments have been developed in the laboratory in order to evaluate the oxidation of organic compounds with ozone during drinking water treatment [54]. These experiments showed that certain pharmaceuticals react quickly with ozone while others show no reaction, depending on their structural characteristics. Diclofenac, tetracyclines, carbamazepine, 17α-ethinylestradiol, and estradiol showed rate constants higher than $10^6\ M^{-1}\ s^{-1}$ (pH 7 at 20 °C). For water treatment conditions (pH 7–8; $[O_3] = 1$ mg/L) half-lives for these compounds are lower than 1 s, indicating the complete transformation of the parent compound during the ozonation process. Compounds with no reactive sites for ozone reaction, with lower rate constants, were more efficiently removed by reaction with OH radicals when AOPs were used, with rate constants about two or three times faster. For instance, iopromine, with an ozonation constant of $<0.8\ M^{-1}\ s^{-1}$, showed a K_{OH} of $3.3 \times 10^9\ M^{-1}\ s^{-1}$ and ibuprofen, which was only oxidized above 31%, increased this percentage to 84% when OH radicals were formed.

Ternes et al. [33] evaluated the elimination of bezafibrate, clofibric acid, diclofenac, carbamazepine, and pirimidone under laboratory and full-scale DWTP conditions. Ozone was shown to be effective in eliminating carbamazepine and diclofenac (97%, 0.5 mg/L ozone dose), bezafibrate and pirimidone were appreciably removed with percentages above 50% (1 mg/L ozone doses), while clofibric acid was poorly removed even at high ozone doses (<40%, 2.5–3.0 mg/L ozone doses).

Oxidation of EDCs by reaction with ozone has also been experimentally evaluated. Estrogen steroids and nonylphenols reacted with ozone under similar conditions to those applied in water treatment systems [46]. Petrovic et al. evaluated the elimination of neutral and acidic nonylphenols in a Spanish DWTP [55, 56]. An efficiency of 87% in the elimination of these compounds and their halogenated by-products under ozone treatment was obtained.

More recently, bench-, pilot-, and full-scale studies have been performed to evaluate the ozone efficiency in the elimination of 36 diverse contaminants, including PPCPs, hormones, and pesticides in the USA [57]. Results showed that all the compounds were removed with percentages higher than 50% except for TCEP, lindane, and musk ketone, which were eliminated with percentages lower than 20%, and atrazine, iopromide, and meprobamate with removal percentages between 20 and 50%.

Regarding the transformation products generated from emerging contaminant precursors during ozonation, little information is found [58]. Ozonation of carbamazepine was studied in a German waterworks [59], with the conclusion that when this compound was present in raw water, two main products were formed, BQM (benzaldehydehydroquinazolineone) and BQD (benzaldehydequinazolinedione). Additionally, some transformation products could be predicted on the basis of known reaction pathways for specific functional groups [50]. For instance, it is known that secondary and tertiary amines react with ozone giving hydroxylamines and amine oxides, respectively [60]. Formation of these hydroxylamines could be problematic, for example in the case of sulfonamides, of which hydroxylamines are related to hypersensitivity reactions [61].

2.2.2 Chlorination

Chlorine is an oxidant used for disinfection of water supplies. Free chlorine (i.e., HOCl and OCl^-) is commonly used in the USA for disinfection and oxidation of inorganic species. One major drawback in chlorination use is the formation of chlorinated organic compounds (mainly trihalomethanes and haloacetic acids) as DBPs, which are classified as carcinogenic and/or mutagenic compounds [62, 63]. Although the oxidation kinetics for organic compounds are lower than those for ozone or chlorine dioxide, it reacts rapidly with phenolic compounds, mainly through the reaction between HOCl and the deprotonated phenolate anion [64]. The sequential addition to the aromatic ring leads to ring cleavage. The reactivity with phenol moieties could explain the transformation of hormones (estradiol, ethynylestradiol, estriol, estrone) and nonylphenol by chlorine, evaluated in laboratory experiments [46].

Some experiments have been performed in order to assess chlorination effects over several emerging compounds at the laboratory scale [65–68]. The fate and occurrence of PPCPs (including musk fragrances), endocrine disruptors, and other organic contaminants were evaluated during simulated drinking water treatment (25 mg/L of Cl_2; contact time 24 h) [47]. Under these conditions gemfibrocil, hydrocodone, carbamazepine, compounds with primary or secondary amines (diclofenac, sulfamethoxazole, trimethoprim), and compounds with phenolic moieties (estradiol, estrone, ethynylestradiol, acetaminophen; oxybenzone, triclosan; bisphenol A) showed high reactivity with chlorine. On the other hand, the least reactive compounds were those that have electron-withdrawing functional groups or no conjugated carbon bonds (atrazine, BHC, DEET, fluoxetine, iopromide, meprobamate, and TCEP). The chlorination efficiency to eliminate ten antibiotics (carbadox, erythromycin, roxithromycin, sulfadimethoxine, sulfamerazine, sulfamethazine, sulfamethizole, sulfamethoxazole, sulfathiozole, and tylosin) was also evaluated on the laboratory scale and in surface waters [69]. The results obtained showed that a significant removal of all these compounds could be expected during free chlorination in most water treatment utilities. For instance, carbadox was completely removed within 1 min of contact time and at a chlorine concentration of 0.1 mg/L, while macrolides were removed above 85% with 2 h of contact time and 1 mg/L of chlorine.

Oxidation of organic contaminants has been also evaluated in full-scale treatments. Chlorination studies performed in different DWTPs in the USA and Canada, to assess the elimination of pharmaceuticals (i.e., clofibric acid, naproxen, ibuprofen, acetaminophen, fluoxetine), steroids (i.e., estrone, 17β-estradiol), and plasticizers (bisphenol A), showed nondetectable concentrations of the target compounds after the chlorination step [19].

More recently, the effect of chlorine residual to eliminate several pharmaceuticals and other organic compounds (POOCs) has been evaluated in drinking waters from the USA. The addition of free chlorine to finished drinking water is a common practice as a distribution system disinfectant residual. Gibs et al. [70] have evaluated the effect of the addition of 1.2 mg/L of free chlorine in a finished drinking water with 98 POOCs. Results showed that 52 POOCs would remain after 10 days, with an unremarkable reduction in their concentrations.

As previously described, chlorine usually produces undesirable chlorination by-products to some extent. The formation, fate, and toxicity of oxidative by-products from pesticides and EDCs/PPCPs has been studied and assessed as of potential concern [53, 71]. The E-screen performed after chlorination of bisphenol A, 17α-estradiol, and 17α-ethynylestradiol showed a reduction in estrogenic activity after extended exposure time (120 min). Nevertheless, all compounds showed a similar estrogenicity trend, with a higher estrogenicity activity registered during the first phases of oxidation probably related to the formation of chlorination by-products [72]. Chlorination of acetaminophen has also been studied showing the formation of two chlorination ring products, chloro-4-acetamidophenol and dichloro-4-acetamidophenol, and two quinoidal oxidation by-products, 1,4-benzoquinone and *N*-acetyl-*p*-benzoquinone imine (NAPQI). These toxic compounds are associated with acetaminophen overdoses in humans with lethal effects [73].

Chlorination of sulfamethoxazole (SMX), a member of the sulfonamide antibacterial class, has also been studied in wastewater and drinking water matrices. Chlorine reacted with the aniline nitrogen giving the halogenation of the aniline moiety, yielding a ring chlorinated product, and with the SMX sulfonamide moiety to yield the formation of 3-amino-5-methylisoxazole and *N*-chloro-*p*-benzoquinoneimine subproducts [74].

2.2.3 Chlorine Dioxide

Chlorine dioxide (ClO_2) is an oxidant used for disinfection of high quality water, such as groundwater or treated surface water. In Europe, it is also used to protect drinking water distribution at residual concentrations (0.05 to 0.1 mg/L), while in the USA it is mainly used for the preoxidation of surface waters. Compared to chlorine, ClO_2 is generally a stronger and faster oxidant, [75] and is more effective for the inactivation of viruses, bacteria, and protozoa (including the cyst of Giardia and the oocysts of Cryptosporidium). Chemically, ClO_2 has been demonstrated to be a very selective oxidant of specific functional groups of organic compounds, such as phenolic moieties, tertiary amino groups, or thiol groups [76]. Additionally, halogenated DBPs are not formed even under suitable conditions [77]. Nevertheless, other DBPs are formed during ClO_2 reaction. Therefore, chlorite is the major reduction product of ClO_2, considered to be a blood poison [61, 78] and regulated by the USEPA at the 1 mg/L level [79].

Due to the oxidant doses used of ClO_2 in drinking water treatment and its specific reactivity, a complete elimination of parent contaminants is not expected. Nevertheless, this treatment could lead to the deactivation of specific functional groups responsible for parent activity. Chlorine dioxide has demonstrated cleavage of one of the N – C bonds of tertiary amines [80], which would mean the loss of a methyl or amino group in macrolide antibiotics leading to a related and expected decrease in pharmacological activity [81].

Oxidation of several pharmaceuticals by ClO_2 was evaluated in samples from a German DWTP [82]. Water samples were collected before ClO_2 treatment and spiked with the selected pharmaceuticals. Then, ClO_2 doses of 0.95 and 11.5 mg/L were added and samples were analyzed after 30 min of contact time. Under these experimental conditions, bezafibrate, carbamazepine, diazepam, and ibuprofen showed no reactivity while diclofenac was completely oxidized and phenazone derivatives and naproxen showed an appreciable reactivity.

2.3 Membrane Separation

In membrane processes, a semipermeable membrane separates contaminants from the water by a process known as crossflow filtration (also called tangential flow filtration). The bulk solution flows over, and parallel to, the filter surface while, under pressure, a portion of the water is forced through the membrane to produce a permeate stream. The turbulent flow of the feedwater over the membrane surface minimizes accumulation of particulate matter there, and facilitates continuous operation.

Different types of membranes are applied to drinking water treatment with different characteristic separations depending on their composition and pores. Several classifications can be made to characterize membranes; size exclusion is one of the most significant mechanisms to separate contaminants.

2.3.1 Ultrafiltration

Ultrafiltration (UF) allows the removal of turbidity, microorganisms, and many hydrophobic macromolecules (0.001–0.1 μm) with $\log K_{ow} > 4$. The removal properties of UF membranes are usually expressed in terms of molecular weight cutoff (MWCO) which ranges from 1000 up to 50 000 Da. Nevertheless, most organic EDC/PPCP compounds range from 150 to 500 Da, and only those associated with particles or colloidal organic matter are removed.

An investigation on the removal of 52 EDCs and PPCPs with different physicochemical properties such as size, hydrophobicity, and acidity by UF and nanofiltration (NF) has been carried out in model and natural waters [104, 105]. The results showed that the UF membrane retained hydrophobic EDCs mainly by adsorption processes. UF membranes showed retention percentages lower than 40% for all compounds except triclosan (87%), oxybenzone (77%), and progesterone (56%). In most cases, the concentration of EDCs and PPCPs was feed > retentate > permeate except for few compounds (i.e., diclofenac, erythromycin, estriol, gemfibrozil, ibuprofen, chlordane, dieldrin) that showed lower concentrations in retentate than initial ones. These compounds were probably adsorbed onto the membrane and into the membrane pores. It has been reported that retention of relatively hydrophobic compounds and hormones (i.e., $\log K_{ow} > 3$) by UF, reverse osmosis (RO), and NF membranes is mainly due to adsorption [106, 107]. Yoon et al. [105] stated that compounds highly retained by UF (30–80%) have common structural properties including aromatic ring structures, high pK_a, and/or high $\log K_{ow}$ values, whereas poorly retained compounds include those with low $\log K_{ow}$ due to aliphatic, aromatic, nitrogen, carbonyl, phosphate, amine, or hydroxyl functional groups.

2.3.2 Nanofiltration/Reverse Osmosis

NF and RO are effective physical diffusion-controlled and size-exclusion processes which have been demonstrated to effectively remove pathogens and organic contaminants. However, the rejection efficiency correlates to different parameters affecting the solute, the membrane, and the feed water composition; moreover, it is also correlated with the concentration of the organic contaminant and its effective charge state. Both processes have the broadest duration of treatment capability but require a great degree of pretreatment, and in addition RO has a high relative cost compared with other technologies.

Bench-scale tests have been performed in order to evaluate the removal of several emerging contaminants by NF and/or RO. A pilot system with RO membranes was used to evaluate the elimination ratio of several pharmaceuticals, pesticides, and PPCPs. The system evaluated both virgin and fouled membranes, showing that target analytes were well-rejected and no effect of membrane fouling was detected [108]. Another study evaluated the elimination of steroid hormones by RO in wastewater matrices. Results showed removals greater than 90% for 17β-estradiol and 17α-ethinylestradiol [66].

NF membranes have also been evaluated by bench-scale tests for the analysis of EDCs and PPCPs [109]. Results showed that NF membranes had a low adsorption capacity for the less volatile and less hydrophobic compounds. Average retention percentages were 30–90% depending on their properties, except for naproxen which showed poor retention lower than 10%. In these tests, hydrophobicity led to adsorption and polarity to charge repulsions that were more important than molecular weight in removing EDCs and PPCPs.

A study of the removal of pesticides [110] and pharmaceuticals [111] by NF and RO membranes in a real DWTP has been performed. The DWTP supplies treated water to 20 000 inhabitants and uses one NF line and two parallel RO lines with a final mixing of the three permeates to obtain treated water. Triazines (i.e., simazine, atrazine, terbutylazine, and terbutryn) and metabolites (DIA, DEA) were fully eliminated in both NF and RO lines. On the other hand, removal of pharmaceuticals showed very similar percentages to those obtained for triazines, and high values above 80% were obtained in both NF and RO lines for most of the selected compounds (i.e., hydrochlorothiazide, ketoprofen, gemfibrozil, diclofenac, sulfamethoxazole, sotalol, metoprolol, propylphenazone, and carbamazepine). However, strong fluctuations in the permeate concentrations for some compounds, such as acetaminophen and mefenamic acid, were measured.

An assessment of removal possibilities with NF of priority pollutants in water sources of Flanders and The Netherlands has been recently reviewed [112]. The authors suggested that rejection of organic pollutants in NF could be qualitatively predicted as a function of a limited set of solute parameters, such as $\log K_{\mathrm{OW}}$, $\mathrm{p}K_{\mathrm{a}}$, and molar mass. The prediction was based on the scheme proposed by Bellona et al. [113] but using hydrophobicity as the primary solute parameter. Their qualitative predictions for target compounds (hormones, industrial chemicals, pesticides, and pharmaceuticals) roughly correlated with values from the literature. The authors stated that the solute parameters together with a knowledge of the membrane material can give real estimations of the rejection of organic micropollutants and can provide feasible evaluations of NF in drinking water plant designs.

3
Emerging Disinfection By-Products

A widely known group of drinking water contaminants are DBPs which are generated during the treatment process. Some of these compounds, such as trihalomethanes, haloacetic acids, bromates, or chlorites, are widely known and they have been studied and regulated for the last 30 years. However, emerging contaminants in raw waters and new alternative disinfectants and treatments for drinking water production, implemented by the DWTPs, could lead to the formation of new DBPs. In Sect. 2, DBP formation from pharmaceuticals and hormones was examined. In this section, the emerging DBPs generated during water treatment due to alternative disinfectants from chlorine (i.e., ozone, chlorine dioxide, and chloramines) will be discussed. Up to now, scarce information about the potential toxicity of DBPs generated from these alternative disinfectants can be found. New DBPs identified include iodo-acids, bromonitromethanes, iodo-trihalomethanes, brominated forms of MX, bromoamines and bromopyrrole [114], nitrosodimethylamine (NDMA), and other nitrosamines. Recent studies [115] of their toxicity have demonstrated that some of these compounds are more genotoxic than many of the DBPs regulated, and are present at similar concentration levels to those regulated.

Among the emerging DBPs investigated, one remarkable compound is NDMA [116–118] which is generated from chloramines or chlorine disinfection (Fig. 1) [128–130]. This compound belongs to the chemical class of the *N*-nitrosoamines and its importance remains, as it is considered a potential human carcinogen with more cancer potencies than those reported for trihalomethanes [119, 120]. In 1989, NDMA was first detected in treated drinking water from Ohsweken (Ontario, Canada) at elevated concentrations (up to 0.3 μg/L). This finding prompted a survey of 145 Ontario DWTPs [116, 121, 122] and the concentrations of NDMA detected in the treated water were lower than 5 ng/L (except for some samples exceeding 9 ng/L). More recently, similar results were obtained for NDMA concentrations in drinking water systems from the USA. Results showed that NDMA was detected at concentra-

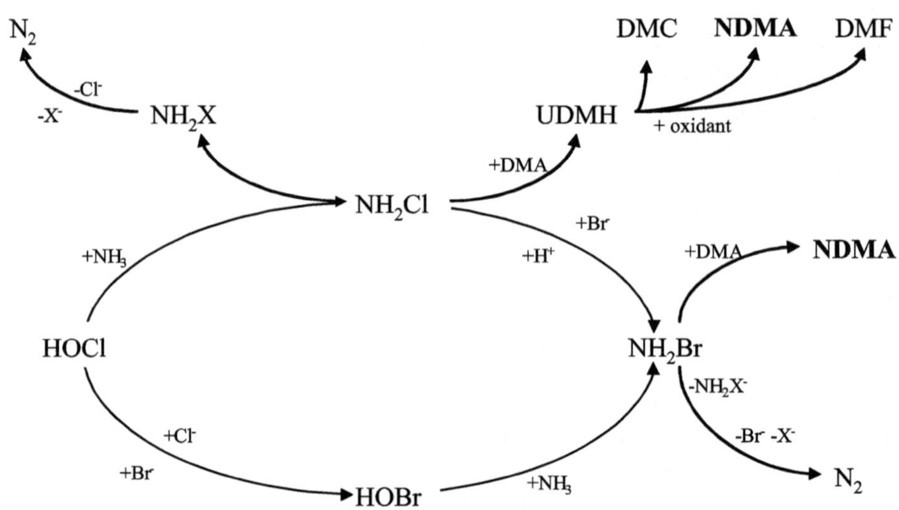

Fig. 1 NDMA formation mechanism for the chloramine/bromamine pathway [128–130]. X: Cl/Br; UDMH: unsymmetrical dimethylhydrazine; DMC: dimethylcyanamide; DMF: dimethylformamide

tion levels lower than 5 ng/L in water supplies which used only free chlorine, while 3 out of 20 chloraminated supplies contained concentrations higher than 10 ng/L [99]. Nevertheless, a more extended survey performed from 2001 to 2002 in 21 USA water systems indicated median concentrations of NDMA in chlorinated or chloraminated waters lower than 1 ng L^{-1} [123]. Regarding legislation, although NDMA is listed as a priority pollutant [124], a maximum contaminant level (MCL) has not been established and it has not yet been included on the candidate contaminant list (CCL), which is a list of unregulated contaminants for monitoring in the USA [125]. Nevertheless, some regulatory agencies have established guidelines for maximum concentrations of NDMA; the Ontario Ministry of the Environment and Energy has fixed a value of 9 ng/L [126], while the California Department of Health Services has suggested a value of 10 ng/L [127].

With regard to other emerging DBPs, Richardson et al. [131] studied the formation of DBPs when alternative disinfectants were used. Over 200 DBPs were identified and a comparison between by-products formed from different treatments was also performed. The effect of high concentrations of bromide on the formation of chlorine dioxide DBPs was also evaluated by selecting natural waters from Israel (Sea of Galilee) with high natural levels of this compound. The DBP structures identified showed high degrees of bromide, such as 1,1,3,3-tetrabromopropane.

Finally, new alternative routes of exposure to drinking water DBPs are now being recognized. Inhalation or dermal absorption during bathing or showering can be translated into high exposure to toxic/carcinogenic compounds [132]. A recent study performed by Villanueva et al. [133] revealed a correlation between these activities and a higher risk of bladder cancer. An additional new route of exposure to DBPs is swimming pools. Zwiener et al. [134] published a review article on the formation of DBPs in swimming pool waters and the adverse health effects that could be related to them.

4 Removal of New Emerging Contaminants in a Drinking Water Treatment Plant (DWTP)

Human habits and activities have been widely demonstrated to impact the environment in many ways. Recently, a new group of human-use contaminants, illicit drugs, have been detected in aquatic media from the USA [135], Italy [136, 137], Germany [138], Spain [139, 140], and Ireland [141]. Due to the high consumption rates—around 200 million people have consumed illicit drugs in the last year—the determination of these compounds has become an important issue, not only for forensic sciences but also in environmental studies [135]. Some of these drugs are released unaltered or as slightly transformed metabolites. Therefore, they reach municipal wastewater treat-

Table 3 Drug concentrations in WWTP samples (NE Spain) (April to September, 2006) and in water from the Llobregat river (NE Spain) (September 2006) [139]

	WWTP influent ($n:16$)			WWTP effluent ($n:16$)			River ($n:6$)		
Compound	Samples (>LOQ[a])	*C* max ng/L	*C* mean ng/L	Samples (>LOQ[a])	*C* max ng/L	*C* mean ng/L	Samples (>LOQ[a])	*C* max ng/L	*C* mean ng/L
Nicotine	10	56053	13082	6	4775	2669	5	815	595
Cotinine	16	6820	2732	12	2726	1419	5	516	331
Caffeine	15	61638	23134	14	22848	4356	6	2991	1926
Paraxanthine	14	54220	14240	12	45681	5932	5	2709	1756
Amphetamine	1	15	15	0	<LOQ[b]	<LOQ[b]	0	<LOQ[b]	<LOQ[b]
MDMA	5	91	49	4	67	41	2	3.5	3
MDEA	1	27	28	0	<LOQ[b]	<LOQ[b]	0	<LOD	<LOD
Ketamine	9	50	41	2	49	19	0	<LOD	<LOD
Cocaine	14	225	79	6	47	17	2	10	6
Benzoylecgonine	14	2307	810	11	928	216	4	111	77

[a] Number of samples with concentrations higher than LOQ value

[b] Concentrations between LOQ and LOD. LOQs (wastewater): nicotine (800 ng/L); cotinine (500 ng/L); caffeine (5 ng/L); paraxanthine (850 ng/L); amphetamine and MDA (1 ng/L); METH (0.9 ng/L); MDMA (1.5 ng/L); MDEA (2.5 ng/L); ketamine (5 ng/L); cocaine and BE (0.2 ng/L). LOQs (surface water): nicotine, cotinine, and paraxanthine (200 ng/L); caffeine (1.5 ng/L); amphetamine, MDA, and MDEA (0.8 ng/L); METH (0.7 ng/L); MDMA (0.3 ng/L); ketamine (3.1 ng/L); cocaine (0.15 ng/L); BE (0.1 ng/L)

ment plants (WWTPs) where, depending on the efficiency of the treatment, they are totally removed or, on the contrary, persist during the treatment and can be detected in receiving waters. The effectiveness of the water treatment processes and the impact of most of these compounds on the aquatic environment are still unknown.

An UPLC-MS/MS method was developed for the analysis of caffeine, nicotine, cocaine, amphetamine related compounds, and other synthetic controlled drugs, and their metabolites, in waste and surface waters [139]. Once the method was optimized and the quality parameters were established, the method was applied to the estimation of the occurrence of these substances in water samples from Catalonia (NE Spain) (Table 3). Results displayed in this table have been already submitted for publication. The analysis of several samples from WWTPs revealed the presence of drugs, such as cocaine and amphetamine related compounds, in both influent and effluent samples. Several illicit drugs, such as cocaine or MDMA (ecstasy), were also found in surface waters while nicotine and caffeine were detected in all the analyzed samples. The results obtained demonstrate that the presence of these drugs in aquatic media must be considered a matter of environmental concern [139].

The incidence of these illicit drugs in surface waters posed the need to investigate the elimination of these compounds during drinking water treatment and their presence in final treated water. The treatment in the DWTP investigated consisted in prechlorination (with chlorine or chlorine dioxide), sand filtration, flocculation and sedimentation, ozonation, GAC filtration, and final postchlorination.

Table 4 Drug concentrations of raw water, treated water, and elimination percentages in a DWTP (Spain)

	Intake[a] ng/L	Treated[a] ng/L	Elimination (%)
Nicotine	nd–1047	<LOQ	>99.9
Cotinine	nd–516	nd–276	74
Caffeine	nd–2991	nd–126	93
Paraxanthine	nd–2709	<LOQ	>99.9
Amphetamine	nd–165	<LOQ	>99.9
MDA	nd–6	<LOQ	>99.9
MDMA	nd–123	<LOQ	>99.9
MDEA	nd–54	<LOQ	>99.9
Ketamine	nd–61	<LOQ	>99.9
Cocaine	nd–411	<LOQ	>99.9
Benzoylecgonine	nd–1047	nd–24	89

nd: non detected
[a] $n = 24$

Several controlled drugs, such as cocaine, benzoylecgonine (cocaine metabolite) and some amphetamine type stimulants (i.e., amphetamine or ecstasy), were detected with concentrations higher than their limit of quantitation (LOQ) at the intake of the selected DWTP [141]. For instance, maximum concentrations of 22 ng/L were obtained for cocaine and up to 37 ng/L for ecstasy. The removal efficiency during treatment was also evaluated and the results (Table 4) showed that removal percentages higher than 99.9% were obtained for most of the compounds found at the intake, including cocaine and ecstasy.

Only three of the studied compounds were detected in some samples with concentrations higher than the LOQs. Cotinine and caffeine among the controlled drugs were found in treated water with removal percentages of about 74 and 93%, respectively, and among the illicit drugs only the biologically inactive metabolite of cocaine was found in treated water at low ng/L levels with a removal of 89%. The analyses were performed by using an UPLC system coupled to tandem mass spectrometry (MS/MS) and the quality parameters were already established [139]. An extracted chromatogram from a treated water sample is displayed in Fig. 2. Two transitions were acquired for each compound in order to obtain four identification points, fulfilling the

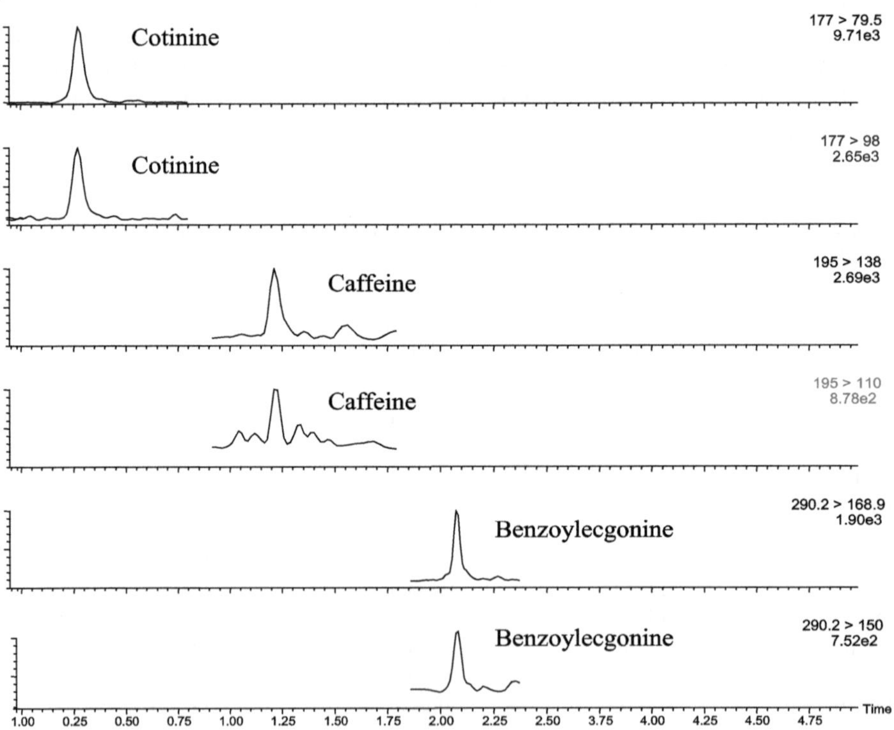

Fig. 2 Extracted ion chromatogram obtained at the intake of a Spanish DWTP. SRM acquisition mode

European Council directives (96/23/EC) regarding mass spectrometric detection [142] and the general criteria for forensic analysis [143].

5
Concluding Remarks

The occurrence of emerging contaminants in aquatic media has been widely assessed in the last decade. Nevertheless, much more data are needed in order to improve the knowledge of the behavior/removal of these compounds in wastewaters and surface waters, as well as their toxicological impact on both aquatic life and human beings and to establish safe guideline values. Moreover, the occurrence of these contaminants in drinking water, just like the removal efficiency of the treatment processes, is still relatively unknown.

In this chapter, a summary of the works published regarding the elimination of emerging contaminants through conventional drinking water treatments and the persistence of some of them through treatments has been presented. Activated carbon adsorption (PAC or GAC) has been shown to be effective to remove nonionic compounds with log K_{ow} higher than 3. Nevertheless, some pharmaceuticals such as carbamazepine and some fragrances such as HHCB (Galaxolide) persisted throughout treatment. NF and RO membranes were also found to remove organic contaminants to a very high extent. Oxidation processes such as ozonation and chlorination have also been evaluated in the elimination of emerging contaminants. Ozone was shown to be very effective in eliminating several pharmaceuticals, hormones, and nonylphenols with percentages higher than 50%, while poorer elimination rates were found for some pesticides (i.e., lindane, atrazine), fragrances (i.e., musk ketone), and pharmaceuticals (i.e., clofibric acid, meprobamate). Oxidation with chlorine or chlorine dioxide was shown to be less efficient but high reactivities were obtained when contaminants contained phenolic or amino moieties (i.e., hormones, nonylphenols, sulfonamides). One major drawback of the oxidation processes is the formation of undesirable DBPs which could have toxic effects. The formation of DBPs from these emerging contaminants together with new disinfection treatments could lead to emerging DBPs. Up to now some new DBPs, such as NDMA, bromonitromethanes, or iodo-trihalomethanes, have already been identified.

Finally, it must be emphasized that the emerging contaminants field is still growing. New human habits or activities could cause the appearance of novel contaminants in aquatic media that may become emerging contaminants. One example of new contaminants derived from human activities and detected in water sources are illicit drugs. These contaminants have recently been detected in aquatic media from the USA and Europe, thus demonstrating once more the cause–effect relationship between human activities and environmental contamination.

References

1. Ternes T (1998) Water Res 32:3245
2. Ternes TA, Kreckel P, Mueller J (1999) Sci Total Environ 225:91
3. Johnson AC, Sumter JP (2001) Environ Sci Technol 35:4697
4. Ternes T (2001) Pharmaceuticals and metabolites as contaminants of the aquatic environment. In: Daughton C, Jones-Lepp T (eds) Pharmaceuticals and personal care products in the environment: scientific and regulatory issues. American Chemical Society, Washington, p 39
5. Ternes T, Stuber J, Herrmann N, McDowell D, Ried A, Kampmann M, Teiser B (2003) Water Res 37:1976
6. Joss A, Keller E, Alder A, Gobel A, McArdell C, Ternes T, Siegrist H (2005) Water Res 39:3139
7. Thomas P, Foster G (2005) Environ Toxicol Chem 24:25
8. Kolpin DW, Furlong ET, Meyer MT, Thurman EM, Zaugg SD, Barber LB, Buxton HT (2002) Environ Sci Technol 36:1202
9. Stumpf M, Ternes T, Wilken R-D, Rodriques S, Baumann W (1999) Sci Total Environ 225:135
10. Snyder S, Kelly K, Grange A, Sovocool GW, Snyder E, Giesy J (2001) Pharmaceuticals and personal care products in the waters of Lake Mead Nevada. In: Daughton C, Jones-Lepp T (eds) Pharmaceuticals and personal care products in the environment: scientific and regulatory issues. American Chemical Society Washington, p 117
11. Barber LB, Leenheer JA, Pereira W, Noyes T, Brown G, Tabor C, Writer J (1995) Contaminants in the Mississippi River from municipal and industrial wastewater. In: Us Geological survey circular 1133, Virginia
12. Singer H, Muler S, Tixier C, Pillonel L (2002) Environ Sci Technol 36:4998
13. Calamari D, Zuccato E, Castiglioni S, Bagnati R, Fanelli R (2003) Environ Sci Technol 37:1241
14. Daughton C (2004) Environ Impact Assess Rev 24:711
15. Zwiener C (2007) Anal Chem 387:1159
16. Heberer T, Stan H-J (1997) Int J Environ Anal Chem 67:113
17. Heberer T, Schmidt-Bäumler K, Stan H-J (1998) Acta Hydrochim Hydrobiol 26:272
18. Loraine G, Pettigrove M (2006) Environ Sci Technol 40:687
19. Boyd GR, Reemstsma H, Grimm DA, Mitra S (2003) Sci Total Environ 311:135
20. Jux U, Baginski RM, Arnold H, Kronke M, Seng PN (2002) Int J Hyg Environ Health 205:393
21. McLachlan JA, Guillette LJ, Iguchi T Jr, Toscano WA Jr (2001) Ann NY Acad Sci 948:153
22. Perret D, Gentili A, Marchese S, Greco A, Curini R (2006) Chromatographia 63:225
23. Williams PRD (2001) Environ Forensics 2:75
24. Achten C, Kolb A, Puttmann W (2002) Environ Sci Technol 36:3662
25. California Department of Health Services. California's experience with perchlorate in drinking water. http://www.drs.ca.gov/ps/ddwem/chemicals/perchl/perchlindx.htm
26. Teixeira G, Costa C, de Carvalho VL, Pereira S, Hage E (1993) Bull Pan Am Health Organ 27:244
27. Burns J (2003) Report to the Florida Department of Health. Tallahassee
28. Stackelberg PE, Furlong ET, Meyer MT, Zaugg SD, Hendersond AK, Reissmand DB (2004) Sci Total Environ 329:99

29. Loos R, Wollgast J, Huber T, Hanke G (2007) Anal Bioanal Chem 387:1469
30. Rabiet M, Togola A, Brissaud F, Seidel JL, Budzinski H, Elbaz-Poulichet F (2006) Environ Sci Technol 40:5282
31. Stackelberg PE, Gibs J, Furlong ET, Meyer MT, Zaugg SD, Lippincott L (2007) Sci Total Environ 377:255
32. Zuccato E, Calamari D, Natangelo M, Fanelli R (2000) Lancet 355:1789
33. Ternes TA, Meisenheimer M, McDowell D, Sacher H, Brauch J, Gulde BH, Preuss G, Wilme U, Seirbet NZ (2002) Sci Total Environ 36:3855
34. Vanderford BJ, Snyder SA (2006) Environ Sci Technol 40:7312
35. Tauber R (2003) Quantitative analysis of pharmaceuticals in drinking water from ten Canadian cities. Enviro-Test Laboratories, Xenos Division, Ontario
36. Reddersen K, Heberer T, Dunnbier U (2002) Chemosphere 49:539
37. Zuehlke S, Duennbier U, Heberer T (2004) J Chromatogr A 1050:201
38. Holger MK, Ballschmiter K (2001) Environ Sci Technol 35:3201
39. Carmichael WW (2001) American Water Works Association Research Foundation. Denver
40. Hoeger SJ, Hitzfeld BC, Dietrich DR (2005) Toxicol Appl Pharmacol 203:231
41. Luks-Betlej K, Popp P, Janoszka B, Paschke H (2001) J Chromatogr A 938:93
42. Psillakis E, Kalogerakis N (2003) J Chromatogr A 999:145
43. Fromme H, Kuchler T, Otto T, Pilz K, Muller J, Wenzel A (2002) Water Res 36:1429
44. Robeck GG, Dostal KA, Cohen JM, Kriessl JF (1965) J Am Water Works Assoc 57:181
45. Sacher F, Haist-Gulde B, Brauch H-J, Preuss G, Wilme U, Zullei-Seibert N, Meisenheimer M, Welsch H, Ternes TA (2000) 219th ACS national meeting, San Francisco, p 116
46. West P (2000) AWWA Annual Conference. Denver
47. Westerhoff P, Yoon Y, Snyder S, Wert E (2005) Environ Sci Technol 36:6649
48. Hoigne J, Bader H (1983) Water Res 17:185
49. Hoigne J, Bader H (1985) Water Res 19:993
50. von Gunten U (2002) Water Res 37:1443
51. Haag WR, Yao CCD (1992) Environ Sci Technol 26:1005
52. Hoigne J (1998) In: Hubrec J (ed) Handbook of environmental chemistry. Springer, Berlin, p 83
53. Acero JL, von Gunten U (2001) J Am Water Works Assoc 93:90
54. Huber MM, Canonica S, Park G-Y, von Gunten U (2003) Environ Sci Technol 37:1016
55. Petrovic M, Diaz A, Ventura F, Barceló D (2001) Anal Chem 73:5886
56. Petrovic M, Diaz A, Ventura F, Barceló D (2003) Environ Sci Technol 37:4442
57. Snyder SA, Wert EC, Rexing DJ, Zegers RE, Drury DD (2006) Ozone Sci Eng 28:445
58. Ikehata K, Naghashkar NJ, El-Din MG (2006) Ozone Sci Eng 28:353
59. McDowell DC (2005) Environ Sci Technol 39:8014
60. Muñoz F, von Sonntag C (2000) Chem Soc Perkin Trans 2:2029
61. Sisson ME, Rieder MJ, Bird IA, Almawi WY (1997) Int J Immunopharmacol 19:299
62. WHO guidelines for drinking-water quality (2004) WHO, Geneva, p 451
63. Miles AM, Singer PC, Ashley DL, Lynberg MC, Langlois PH, Nuckols JR (2002) Environ Sci Technol 36:1692
64. Faust BC, Hoigne J (1987) Environ Sci Technol 21:957
65. Adams SM, Greeley MS (2000) Water Air Soil Pollut 123:103
66. Huang C, Sedlak DL (2001) Environ Toxicol Chem 20:133
67. Sedlak DL, Pinkston KE (2001) Water Res Update 120:56
68. Gould JP, Richards JT (1984) Water Res 18:1001

69. Chamberlain E, Adams C (2006) Water Res 40:2517
70. Gibs J, Stackelberg PE, Furlong ET, Meyer MT, Zaugg SD, Lippincott RL (2007) Sci Total Environ 373:240
71. Huber MM, Ternes TA, von Gunten U (2004) Environ Sci Technol 38:5177
72. Alum A, Yoon Y, Westerhoff P, Abbaszadegan M (2004) Environ Toxicol 19:257
73. Bredner M, MacCrehan WA (2006) Environ Sci Technol 40:516
74. Dodd M, Huang C (2004) Environ Sci Technol 38:5607
75. Ravacha C, Blits R (1984) Water Res 19:1273
76. Hoigne J, Bader H (1994) Water Res 28:45
77. USEPA: Stage 1 disinfectants and disinfection by-product rule (1998). EPA 63-FR 69390–69476
78. Condie LW (1986) J Am Water Works Assoc 73:78
79. USEPA: Alternative disinfectants and oxidants guidance manual (1999). EPA 815-R-99–014
80. Rosenblatt DH, Hull LA, De Luca DC, Davis GT, Weglein RC, Williams HKR (1967) J Am Chem Soc 89:1158
81. Li XQ, Zhong DF, Huang HH, Wu SD (2001) Acta Pharmacol Sin 22:469
82. Huber MM, Korhonen S, Ternes TA, von Gunten U (2005) Water Res 39:3607
83. Arslan-Alaton I, Dogruel S (2004) J Hazard Mater 112:105
84. Andreozzi R, Canterino M, Marotta R, Paxeus N (2005) J Hazard Mater 122:243
85. Lopez A, Bozzi A, Mascolo G, Ciannarella R, Passino R (2002) Ann Chim 92:41
86. Bozzi A, Lopez A, Mascolo G, Tiravanti G (2002) Water Sci Technol Water Suppl 2:19
87. Addamo M, Augugliaro V, Di Paola A, García-López E, Loddo V, Marcí G, Palmisano L (2005) J Appl Electrochem 35:765
88. Calza P, Pazzi M, Medana C, Baiocchi C, Pelizzetti E (2004) J Pharm Biomed Anal 35:9
89. Adams C, Wang Y, Loftin K, Meyer M (2002) J Environ Eng ASCE 128:253
90. Dodd MC, Huang CH (2007) Water Res 41:647
91. Calza P, Medana C, Pazzi M, Baiocchi C, Pelizzetti E (2004) Appl Catal B Environ 53:63
92. Andreozzi R, Caprio V, Marotta R, Vogna D (2003) Water Res 37:993
93. Vogna D, Marotta R, Napolitano A, Andreozzi R, d'Ischia M (2004) Water Res 38:414
94. Doll TE, Frimmel FH (2005) Water Res 39:847
95. Vogna D, Marotta R, Andreozzi R, Napolitano A, d'Ischia M (2004) Chemosphere 54:497
96. Pérez-Estrada LA, Maldonado MI, Gernjak W, Agüera A, Fernández-Alba AR, Ballesteros MM, Malato S (2005) Catal Today 101:219
97. Vogna D, Marotta R, Napolitano A, d'Ischia M (2002) J Org Chem 67:6143
98. Andreozzi R, Caprio V, Marotta R, Vogna D (2003) Water Res 37:993
99. Brillas E, Sirés I, Arias C, Cabot PL, Centellas F, Rodríguez RM, Garrido JA (2005) Chemosphere 58:399
100. Zbaida S, Kariv R, Fischer P, Silmangreenspan J, Tashma Z (1986) Eur J Biochem 154:603
101. Addamo M, Augugliaro V, Di Paola A, García-López E, Loddo V, Marcí G, Palmisano L (2005) J Appl Electrochem 35:765
102. Ohko Y, Iuchi KI, Niwa C, Tatsuma T, Nakashima T, Iguchi T, Kubota Y, Fujishima A (2002) Environ Sci Technol 36:4175
103. Nakamura H, Kuruto-Niwa R, Uchida M, Terao Y (2007) Chemosphere 66:144
104. Feng XH, Ding SM, Tu JF, Wu F, Deng NS (2005) Sci Total Environ 345:229

105. Yoon Y, Westerhoff P, Snyder SA, Wert EC (2006) J Memb Sci 270:88
106. Kimura K, Amy G, Drewes J, Watanabe Y (2003) J Memb Sci 221:89
107. Nghiem LD, Schaeffer AI, Elimelech M (2004) Environ Sci Technol 38:1888
108. Snyder SA, Adham S, Redding AM, Cannon FS, Decarolish J, Oppenheimer J, Wert EC, Yoon Y (2006) Desalination 202:156
109. Yoon Y, Westerhoff P, Snyder SA, Wert EC, Yoon J (2006) Desalination 202:16
110. Quintana J, Ventura F, Martí I, Luque F (2005) EMCO workshop, Dubrovnik
111. Radjenovic J, Petrovic M, Ventura F, Barceló D (2007) EMCO workshop, Belgrade
112. Verliefde A, Cornelissen E, Amy G, Van der Bruggen B, van Dijk H (2007) Environ Pollut 146:281
113. Bellona C, Drewes JE, Xu P, Amy G (2004) Water Res 38:2795
114. Krasner SW, Weinberg HS, Richardson SD, Pastor SJ, Chinn R, Sclimenti MJ, Onstad GD, Thruston AD Jr (2006) Environ Sci Technol 40:7175
115. Richardson SD, Thruston AD Jr, Rav-Acha C, Groisman L, Popilevsky I, Juraev O, Glezer V, McKague AB, Plewa MJ, Wagner ED (2003) Environ Sci Technol 37:3782
116. Jobb DB, Hunsinger RB, Meresz O, Taguchi VY (1995) Proc Am Water Works Assoc, water quality technology conference, Denver
117. Graham JE, Meresz O, Farquhar GJ, Andrews SA (1995) Proc Am Water Works Assoc, water quality technology conference, Denver
118. Najm I, Trussell RR (2001) J Am Water Works Assoc 93:92
119. Mitch WA, Sharp JO, Trussell RR, Valentine RL, Alvarez-Cohen L, Sedlak DL (2003) Environ Eng Sci 20:389
120. US EPA (2002) Integrated risk information system. Office of Research and Development (ORD), National Center for Environmental Assessment, www.epa.gov/ngispgm3/iris/search.htm
121. MOE (1998) Ontario Ministry of the Environment. Drinking Water Surveillance Program, 1996–1997. Executive Summary Report, www.ene.gov.on.ca/envision/dwsp/index96_97.htm
122. DHS (2002) California Department of Health Services; NDMA in California drinking water. March 15, www.dhs.ca.gov/ps/ddwem/chemicals/NDMA/history.htm
123. Barrett S, Hwang C, Guo Y, Andrews SA, Valentine R (2003) Proceedings of the 2003 AWWA annual conference, Anaheim
124. CFR (2001) Code of Federal Regulations, Title 40, Chapter 1, Part 131.36
125. US EPA (1998) Announcement of drinking water candidate contaminant list. Fed Reg 63(40):10273
126. MOE (2000) Ontario Ministry of the Environment and Energy. Regulation made under the Ontario Water Resources Act: Drinking Water Protection—Larger Water Works, www.ene.gov.on.ca/envision/WaterReg/Reg-final.pdf
127. www.epa.gov/safewater/mdbp/dbp1.html
128. www.valleywater.org/media/pdf/SFPUC_NDMA_White_Paper.pdf
129. Mitch WA, Sedlak DL (2002) Environ Sci Technol 36:588
130. Mitch WA, Sharp JO, Trussell RR, Valentine RL, Alvarez-Cohen L, Sedlak DL (2003) Environ Eng Sci 20:389
131. Richardson SD, Thruston AD, Caughran T, Chen PH, Collette TW, Schenck KM, Lykins BW Jr, Rav-Acha C, Glezer V (2000) Water Air Soil Pollut 123:95
132. Richardson S (2007) Anal Chem 79:4295
133. Villanueva CM, Cantor KP, Grimalt JO, Malats N, Silverman D, Tardon A, Garcia-Closas R, Serra C, Carrato A, Castano-Vinyals G, Marcos R, Rothman N, Real FX, Dosemeci M, Kogevinas M (2007) Am J Epidemiol 165:148

134. Zwiener C, Richardson SD, DeMarini DM, Grummt T, Glauner T, Frimmel FH (2007) Environ Sci Technol 41:363
135. Jones-Lepp TL, Alvarez DA, Petty JD, Huckins JN (2004) Arch Environ Contam Toxicol 47:427
136. Zuccato E, Chiabrando C, Castiglioni S, Calamari D, Bagnati R, Schiarea S, Fanelli R (2005) Environ Health 4:1
137. Castiglioni S, Zuccato E, Crisci E, Chiabrando C, Fanelli R, Bagnati R (2006) Anal Chem 78:8421
138. Hummel D, Loffer D, Fink G, Ternes TA (2006) Environ Sci Technol 40:7321
139. Huerta-Fontela M, Galceran MT, Ventura F (2007) Anal Chem 79:3821
140. Boleda MT, Galceran MT, Ventura F (2007) J Chromatogr A 115:38
141. Bones J, Thomas KV, Pull B (2007) J Environ Monit 9:701
142. Commission of the European Communities Official Journal of the European Communities (2002) p 221
143. Rivier L (2003) Anal Chim Acta 492:69

Hdb Env Chem Vol. 5, Part S/1 (2008): 169–188
DOI 10.1007/698_5_107

Published online: 5 April 2008

Impact of Emergent Contaminants in the Environment: Environmental Risk Assessment

Julián Blasco[1] (✉) · Angel DelValls[2]

[1]Instituto de Ciencias Marinas de Andalucía (CSIC), Campus Río San Pedro, 11510 Puerto Real (Cádiz), Spain
julian.blasco@icman.csic.es

[2]Departamento Química-Física, Facultad de Ciencias del Mar y Ambientales, Universidad de Cádiz, Campus Río San Pedro, 11510 Puerto Real (Cádiz), Spain

Abstract Human pharmaceuticals enter the environment mainly through regular domestic use. Their presence in the aquatic environment has been recorded in the range ng L^{-1} to μg L^{-1}. Knowledge of the risk associated with the use of pharmaceuticals involves establishing the ratio between predicted environmental concentrations (PECs) and predicted no effect concentration (PNECs). The European Union (EMEA) and USA (FDA) have implemented two-tiered strategies for environmental risk assessment (ERA) of pharmaceuticals. Advances in analytical techniques have allowed us to measure pharmaceuticals in the environmental compartment and the refinement of ERA. On the other hand, for calculation of PNECs, acute and chronic toxicity tests are employed; a critical analysis of the available information was carried out, indicating that acute toxicity was only likely for spills, although an exception to this general behavior is shown by endocrine-active substances. Studies including mixtures of pharmaceuticals are not common in the study of pharmaceutical effects. Only for a limited number of drugs, are the ecotoxicity data available adequate for risk assessment. Selection of model compounds with a priori knowledge about the target biological compounds, and the selection of species, life stages and endpoints would be helpful. New technologies such as proteomics and genomics could be valuable resources to be included in the framework of pharmaceutical environmental risk assessment.

Keywords Ecotoxicology · Environmental concentration · Pharmaceuticals · Risk assessment

Abbreviations

AF	Assessment factor
BAF	Bioaccumulation factor
CPMP	Committee for Proprietary Medicinal Products
EC_{50}	Effect concentration 50%
EE2	Ethynilestradiol
EEC	Expected environmental concentration
EIC	Expected introduction concentration
EMC	Endocrine modulating chemicals
EMEA	European Medicines Agency
ERMS	European Risk Management Strategy
FDA	Food and Drug Administration
GMOs	Genetically modified organisms
ICH	International Conference on Harmonization of Pharmaceuticals for Human Use
ISO	International Organization for Standardization
LC_{50}	Lethal concentration 50%
LC-MS	Liquid chromatography tandem mass spectrometry
LOQ	Limit of quantification
NOEC	No observed effect concentration
OA	Oxolonic acid
OECD	Organization for Economic and Cooperation Development
OTC	Oxytetracycline
PBDEs	Polybromated diphenylethers
PEC	Predicted environmental concentration
PNEC	Predicted no effect concentration
PPCPs	Pharmaceutical and personal care products
QSARs	Quantitative structure—activity relationships
SSRI	Selective serotonin re-uptake inhibitors
STP	Sewage treatment plant
TGD	Technical Guide Document in Support of Commission Directive 93/67/EEC

1 Introduction

Emergent contaminants are not easy to define because they represent a changing reality, dependent on perspective and timing [1]. The permanence in this status is dependent on its persistence in the environment, effects on humans and ecotoxicity. In this sense, knowledge of new properties of chemicals that are well known can re-introduce them as emergent contaminants. Recently, an editorial of Environmental Toxicology and Chemistry [2] pointed out that the level of concern about the new emergent contaminants is unknown and it is necessary to evaluate their significance for human and ecological health.

Four broad categories have been established for emergent contaminants: (a) pharmaceuticals and personal-care products (PPCPs); (b) polybromated diphenylethers (PBDEs) and other persistent organic contaminants; (c) endocrine modulating chemicals (EMCs) and (d) nanotechnology products. These categories are not totally separated because a compound could be at the same time a PPCP and an EMC.

Herein we will focus on the environmental risk assessment of human pharmaceuticals because the ERA of the different types of emergent contaminants pointed out above is beyond the scope of this work.

Entry of human pharmaceuticals and PCCPs to the environment is mainly via regular domestic use [3]. After their use, pharmaceuticals are excreted, some of them are partially metabolized (slightly transformed or conjugated to polar molecules) and released into the aquatic environment via wastewater effluent. Unused drugs are stored until the expiration date and finally exposed of down drains reaching the aquatic environment. Consequently, they can potentially affect drinking water quality. The entry path scenarios for human pharmaceutical products have been summarized by the Committee for Proprietary Medicinal Products (CPMP) (Fig. 1) [4].

Variable quantities of pharmaceuticals are present in surface waters, ground waters, and sediment, ranging in concentrations between ng L^{-1} to μg L^{-1} [5, 6]. Knowledge of pharmaceuticals in environmental compartments has been supported by the great advance in analytical techniques, which has improved detection levels of these compounds in the environment. New chemical methods, such as liquid chromatography tandem mass

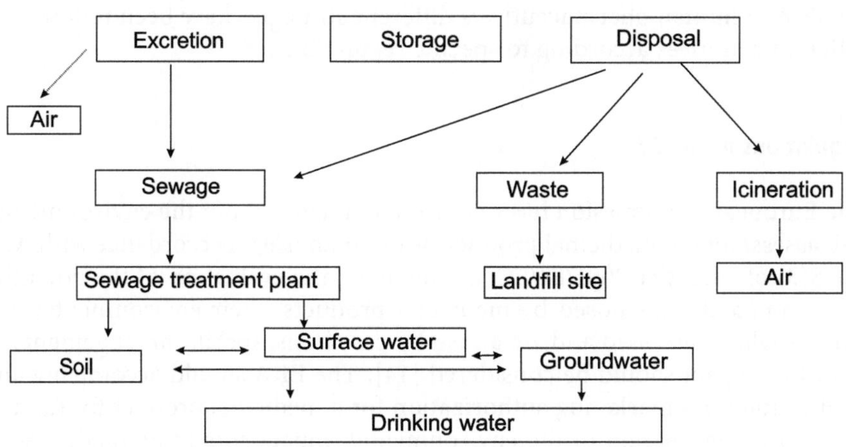

Fig. 1 Routes of entry to the environment for human pharmaceuticals [4]

spectrometry (LC-MS), are able to determine more organic polar compounds without derivatization [7–9]. As a consequence, several monitoring programs have been carried out in different countries that have demonstrated the presence of drug residues to be widely distributed.

On the other hand, knowledge concerning the ecotoxicological effects of pharmaceuticals on aquatic and terrestrial organisms and wildlife is scarce, especially the aspects related to chronic toxicity and more-subtle effects [10]. Most of the published aquatic toxicity data and risk assessments for human pharmaceuticals are based on short-term acute studies [5, 11, 12]. Nevertheless, information about the chronic effects of human pharmaceuticals on aquatic organisms has been recently reviewed by Crane et al. [13].

Although the amounts of human drugs released to the environment are quite high, only recently have detailed guidelines been developed about how pharmaceuticals should be assessed.

2 Environmental Risk Assessment Regulations

Environmental risk assessment is a process that evaluates the likelihood that adverse effects may occur as a result of exposure to one or more stressors [15]. The characterization of the risk involves knowing the ratio between predicted environmental concentration (PEC) and predicted no effect concentration (PNEC); if this value is less than 1 there is no risk to the ecosystem, but if the value is equal to or higher than 1 there is a risk and regulation activities will be needed.

Although the market for pharmaceuticals is highly globalized, and harmonization for testing guidelines have been supported by the International Conference on Harmonization of Pharmaceuticals for Human Use (ICH), for the ERA of human pharmaceuticals different strategies have been followed in different countries according to specific regulations.

2.1 Regulations in the EU

The European Commission has released a guideline about the environmental risk assessment of medicinal products for human use, in accordance with Article 8(3) of Directive 2001/83/EC, as amended, the evaluation of the potential environmental risks posed by medicinal products, their environmental impact should be assessed and, on a case-by-case basis, specific arrangements to limit the impact should be considered [14]. The ERA should accompany any application for a marketing authorization for a medicinal product for human use and the evaluation of the environmental impact should be made also if there is an increase in the environmental exposure. Nevertheless, this guideline does not apply to medicinal products consisting of genetically modified organisms (GMOs).

The evaluation of risk assessment to the environment is a step-wise process, consisting of two phases. The first phase (Phase I) includes checking the exposure of the environment to the drug substance against the action limit assessment. If the result is lower than the limit assessment the ERA is finished. Alternatively, second-phase information about the fate and effect of the drug substance should be carried out. This Phase II is divided into two parts (Tier A and B). In Table 1, the phase approach of environmental risk assessment according to the guidelines of EMEA is shown [14]. Phase I is considered a pre-screening and it is independent of route administration, pharmaceutical characteristics, metabolism, and excretion. The calculation of PEC is restricted to the aquatic environment and some restrictions are considered:

- A market penetration factor (Fpen) is defined, the value can be a default value or refined according to specific data (eg. Epidemiological data).
- The amount is distributed along the year and the considered geographic area.
- The sewage system is the main route of entry for the substances.
- No biodegradation of the substance is taken into account during the treatment in the sewage treatment plant (STP).
- Metabolism in the patient is not considered.

For calculation of the PEC the following equation is applied [14]:

$$PEC_{surfacewater} = \frac{Dose_{ai} \times F_{pen}}{Wastewater_{inh} \times Dilution} \quad (1)$$

Table 1 The phase approach in environmental risk assessment according to the Committee for Medicinal Products for Human Use [14]

Stage in regulatory evaluation	Stage in risk assessment	Objective	Method	Test/data requirement
Phase I	Pre-screening	Estimation of exposure	Action limit	Consumption data, log K_{ow}
Phase II Tier A	Screening	Initial prediction of risk	Risk assessment	Base set aquatic toxicology and fate
Phase II Tier B	Extended	Substance and compartment – specific refinement and risk assessment	Risk assessment	Extended data set on emission, fate and effects

where $Dose_{ai}$ ($mg\,inh^{-1}\,d^{-1}$) is the maximum daily dose consumed per inhabitant; F_{pen} is the percentage of market penetration and represents the proportion of the population being treated daily with a specific substance; $Wastewater_{inh}$ ($L\,inh^{-1}\,d^{-1}$) corresponds to the amount of wastewater per inhabitant and per day and Dilution is the dilution factor.

When the $PEC_{surfacewater}$ value is below 0.01 $\mu g\,L^{-1}$ and there are no other environmental concerns it is assumed that the pharmaceutical is not a risk. In the case where the $PEC_{surfacewater}$ is above this value, a Phase II environmental fate and effect analysis should be carried out. In drugs that have a $PEC_{surfacewater}$ lower than 0.01 $\mu g\,L^{-1}$ but may affect reproduction a strategy including Phase II evaluation should be carried out.

In the Phase II assessment, the evaluation of the PEC/PNEC ratio is based on aquatic toxicology data and predicted environmental concentration (Tier A). For drugs where a potential impact can be weighted a refinement of the values should be realized in Tier B. The guidelines for experimental bioassays of the Organization for Economic Cooperation and Development (OECD) or the International Organization for Standardization (ISO) should be followed and all relevant data about physical-chemical properties, metabolism, excretion, biodegradability, persistence, and pharmacodynamic processes must be taken into account.

For the aquatic effect analysis standard long-term toxicity tests in fish, *daphnia*, and algae are proposed (OECD 201, 210, and 211) [16] and to determine the $PNEC_{water}$ an assessment factor (AF) is applied to the no-observed effect concentration (NOEC).The AF applied is a default value of 10 and it represents the uncertainty associated to intra-species variability and inter-species sensitivities and extrapolation from lab to field studies.

The refinement of the risk when it has been identified in Tier A involves refining PEC and PNEC values for the compounds using data on transformation of the substance in the environment. The equation that should be applied is:

$$PEC_{surfacewater} = \frac{Elocal_{water} \times F_{stpwater}}{Waste_{inh} \times Capacity_{stp} \times Factor \times Dilution} \quad (2)$$

$$Elocal_{water} = Dose_{ai} \times F_{excreta} \times F_{pen} \times Capacity_{stp} \quad (3)$$

$Waste_{inh}$ = amount of wastewater per inhabitant per day

$Capacity_{stp}$ = capacity of local sewage treatment plant

$F_{stpwater}$ = fraction of emission directed to surface water

Factor = factor to take into account the adsorption to suspended matter

Dilution = dilution factor

$Elocal_{water}$ = local emission to wastewater of the relevant residue.

If the pharmaceuticals can be adsorbed on soil or sediment, an effect analysis on sediment-dwelling organisms should be carried out and compared

Table 2 Terrestrial fate and effects studies recommended in Phase II Tier B, according to the Committee for Medicinal Products for Human Use [14]

Study type	Recommended protocol
Aerobic and anaerobic transformation in soil	OECD 307
Soil microorganisms: Nitrogen transformation test	OECD 216
Terrestrial plants, Growth test	OECD 208
Earthworm, Acute toxicity tests	OECD 207
Collembola, Reproduction test	ISO 11267

to $PEC_{sediment}$ (OCDE 308) [16]. For compounds with $K_{OC} > 10\,000\,L\,kg^{-1}$, unless they are readily biodegradable, methodologies such as TGD [17] are recommended for risk assessment including PEC_{soil} calculation. The bioassays recommended for Phase II Tier B in soils are shown in Table 2.

Recently, the European Risk Management Strategy (ERMS) work programme for 2008 and 2009 has been adopted, which will focus on improvement of the EU Pharmacovigilance system and the science and methodologies which give support to the safety monitoring of medicines for human use [17].

2.2 Regulations in USA

The National Environmental Policy Act of 1969 requires the Food and Drug Administration (FDA) to take into account the environmental impact of approving drug and biologic applications as an integral part of its regulatory process. A guidance was prepared by the direction of the Chemistry Manufacturing Controls Coordinating Committee, Center for Drug Evaluation and Research (CDER) and the Center for Biologics Evaluation and Research (CBER) and it represents the current thinking on environmental assessment. This guidance [18] involves several topics, among them: the content and format of environmental assessment (EAs), test methods and specific guidance for the environmental issues that are associated with human drugs.

According to this guidance, the EA is required when the estimated concentration of the compound is: (a) equal or higher than $1\,\mu g\,L^{-1}$; (b) when the substance occurs naturally but its application alters significantly its concentration or distribution or its metabolites and (c) when the expected exposure levels can potentially generate harm to the environment. A tiered approach is employed to assess the environmental fate and effects of pharmaceuticals (Fig. 2).

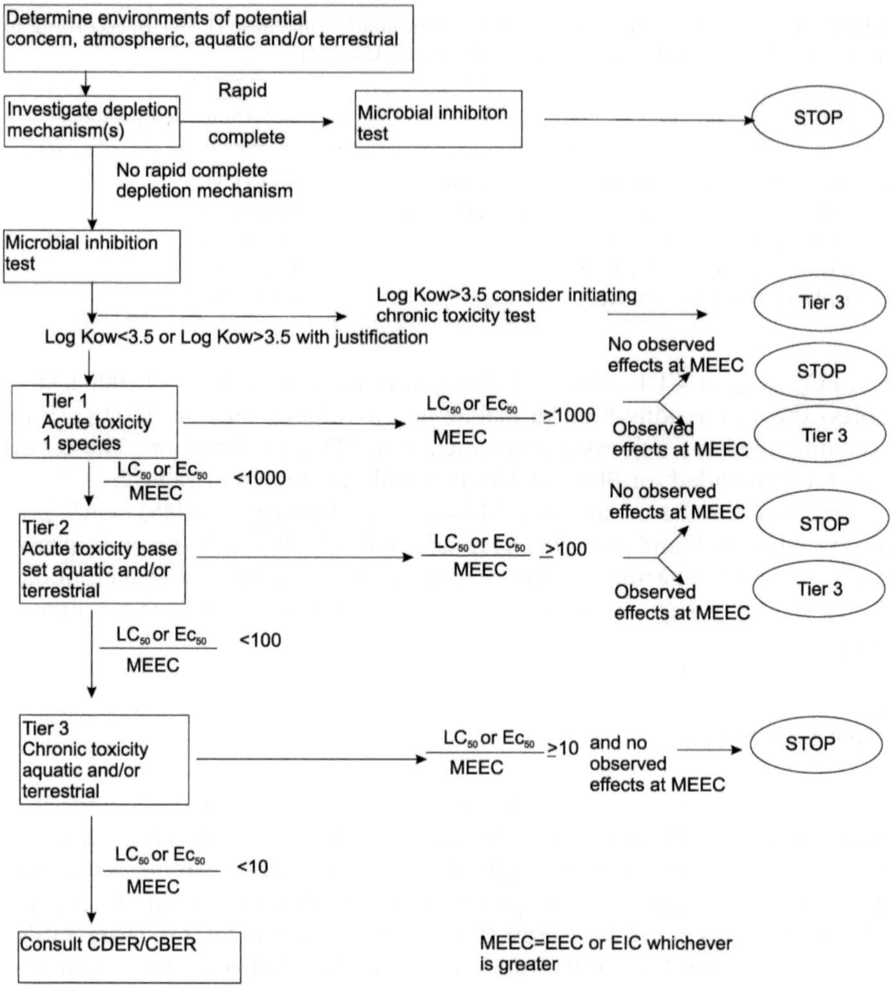

Fig. 2 Tiered approach of FDA for fate and testing [18]

The expected introduction concentration (EIC) should be estimated and the method for calculating this value in aquatic media is:

$$\text{EIC} - \text{aquatic(ppb)} = A \times B \times C \times D$$

$A = \text{kg y}^{-1}$ produced for direct use
$B = 1/\text{L}$ per day entering in STP
$C = \text{year}/365$ days
$D = 10^9\ \mu\text{g kg}^{-1}$

Some kinds of drug may enter the terrestrial environment when biosolids from waste water treatment plant facilities with adsorbed material are applied to soil. The calculation of this concentration is carried out considering the typical treatment, disposal, and application processes. A metabolizing process (biodegradation) occurs during the waste treatment process and it should be considered for calculating EIC.

The PEC is calculated using EIC and taken into account are the processes which affect the compound (spatial or temporal variations, dilution, degradation, sorption, etc.). Normally, EPA applies a dilution factor of 10 to the EIC-aquatic to estimate the PEC.

In summary, the fate of the substance should be provided for the environmental compartment and the transport between compartments should be taken into account if it is of interest to the environmental behavior of the compound.

The evaluation of the effect of pharmaceuticals is oriented to the aquatic compartment because their effect will be on aquatic organisms. Nevertheless, for compounds with high adsorption capacity or high degradation rate, its effects in the aquatic environment could not be considered. For the terrestrial environment, fate and effects testing should be considered when the substance has a $K_{OC} > 10^3$.

Testing of the environmental effects of the pharmaceuticals should be carried out according to the tiered approach as was indicated in Fig. 2. If the compound is not removed from the environment quickly, its persistence and the associated toxic effects should be taken into account. A tiered approach should be used (as was proposed in the guidance), thus the ratio between LC_{50} or EC_{50} and the EIC or EEC is employed as the assessment factor (10, 100, and 1000) to carry out toxicity tests at different levels. The toxicity tests should be performed according to the protocols defined by FDA, OECD, and other peer-reviewed literature if they are appropriate for environmental studies.

3 Pharmaceutical Environmental Concentrations

3.1 Predicted Environmental Concentration

The ERA requires one to know the occurrence and concentration of compounds in the environmental compartments. The exposure assessment should take into account the fate of the substance released to the environment and predict the environmental concentration [19]. The lack of information about measured levels of pharmaceuticals in environmental compartments mean that to carry out the ERA for pharmaceuticals the $PECs_{surfacewater}$

have been estimated, in many cases, according to the recommendations of EMEA or FDA [14, 18]. A review of 111 substances, corresponding to the highest-selling human drugs that have annual sales in Germany of more than 5000 kg, has been carried out. For all compounds the values were higher than 0.01 $\mu g\,L^{-1}$ [20]. According to the scheme developed by EMEA a Phase II process should be carried out for evaluating the exposure. The $PEC_{surfacewater}$ for pharmaceuticals according to data for its use in Germany, Sweden, France and UK [19–23] are presented in Fig. 3. The differences among $PEC_{surfacewater}$ should be related to drug prescription patterns in the countries. These data correspond to the worst case because degradability is not considered. Thus, for paracetamol the PEC is 367.3 $\mu g\,L^{-1}$ [19], although a high degree of elimination, around 98%, has been observed during activated-sludge wastewater treatment [7]. On the other hand, for other compounds such as oxytetracycline (OTC), human metabolism is limited [24], and the compound will be excreted without transformation. It has been observed that biodegradation

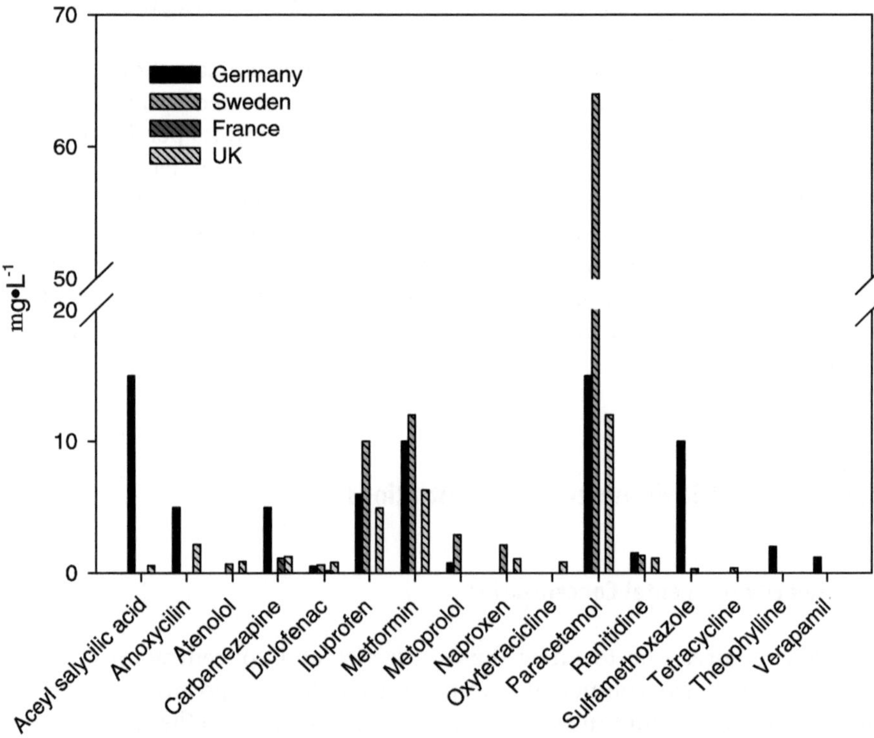

Fig. 3 Predicted environmental concentration (PEC) for pharmaceuticals in surface water of several countries (Germany, Sweden, France, and UK). Data were extracted from [20, 21, 23, 57]

for OTC is limited [25]; and its PEC will be equal to 0.62 $\mu g\,L^{-1}$ after applying a dilution factor of 10.

3.2 Measured Environmental Concentration

3.2.1 Effluent Sewage Treatment Plant

The first work on the presence of drug residues in STP effluents was carried out in the USA and it was focused on clofibric acid, the metabolite of three lipid regulators: clofibrate, etofyllin clofibrate and etofibrate at $\mu g\,L^{-1}$ concentration levels in treated sewage [26]. Later, significant advances in analytical techniques have allowed one to measure pharmaceuticals in environmental compartments [27]. The main drawback of the conventional analytical approach is target-compound monitoring which is insufficient to assess the environmental relevance of emerging contaminants, and the lack of knowledge about the transformation products. Other problems relating to conjugated metabolites (e.g. glucuronides and sulfate conjugates) which can be deconjugated by microbial actions in STP have been pointed out [28].

The pharmaceutical levels in the effluents of STP in many countries are high. Table 3 presents information on the levels for individual compounds in the effluents of STP in Germany, Greece, Spain, and Switzerland. The highest concentrations were recorded in the effluent of STP in Seville (Spain) for two anti-inflammatory drugs, ibuprofen and naproxen, with concentrations of 48.2 and 4.3 $\mu g\,L^{-1}$, respectively [29]. The differences between influent and effluent showed the degradability of these compounds. The values recorded for ibuprofen in the Seville STPs are very high, because the concentrations are below 1 $\mu g\,L^{-1}$, normally. Acetylsalicylic can be degraded into its metabolites, although they are eliminated in the STP process; thus only the metabolite salicylic acid has been detected in sewage effluents [30, 31]. The ubiquity of target compounds can be related to the metabolism, sales, and practices carried out in each country. Therefore, analgesics and antibiotics are detected frequently because they are excreted as the unchanged parent compound; in addition the high loads of analgesic and anti-inflammatories, in comparison with other therapeutic groups is attributed to the higher consumption. The removal efficiency is related to the treatment applied in each plan and the compound physicochemical characteristics and hydraulic retention time [32].

3.2.2 Environmental Levels

In developed countries, production and use of pharmaceuticals are increasing annually [33]. The measurement of these compounds in environmental

Table 3 Concentration range and mean concentration in μg L^{-1} of pharmaceuticals and metabolites in effluents of municipal STPs of several countries

Drug	Germany	Greece	Spain	Switzerland	Canada
Acetyl salycilic acid	0.32–0.92	na	na	na	na
Diclofenac	0.21–1.11	0.20–0.34	blq–0.38	0.1–0.7	0.015–0.039
Ibuprofen	0.32–0.58	na	0.78–48.24	0.005–1.5	2.2–3.5
Naproxen	0.12–0.53	nd	0.22–4.28	0.1–3.5	1.0–1.7
Indometazine	0.07–0.11	na	na	na	0.048–0.075
Benzafibrate	0.72–1.2	nd–0.15	na	na	0.13–0.28
Gemfribozil	0.12–0.35	na	na	na	0.37–0.60
Fenofibric acid	0.32–0.44	nd	na	na	na
Clofibric acid	0.42–0.69	na	na	nd–0.06	na
Carmabezapine	1.31–2.2	na	blq–1.29	0.1–0.8	na
Phenazone	0.12–0.20	na	na	na	na
Porpanolol	0.34–0.48	na	na	na	na
Metoprolol	1.72–2.44	na	na	na	na
Bisoprolol	0.12–0.16	na	na	na	na
Betaxolol	0.14–0.20	na	na	na	na
Terbutalin	0.10–0.12	na	na	na	na
Carazolol	0.05–0.09	na	na	na	na
Dihydrocodeine	1.47	na	na	na	na
Hydrocodone	0.72	na	na	na	na
Ketoprofen	nd	0.27–0.82	blq–3.48	nd–0.20	0.015
Mefenamic acid	nd	0.08–0.22	na	na	na
Primidone	nd–0.88	nd	na	na	na
Propyphenazone	nd–0.74	nd	na	na	na
Salycilic acid	nd–0.65	0.64–2.0	0.57	na	0.054–0.46
Caffeine	na	na	0.15–3.20	na	na

* Data were extracted from [6, 29, 36]
na not analysed, *nd* not detected, *blq* below limit quantification

compartments can improve knowledge about the occurrence and persistence of the compounds in the environment. The advances in analytical techniques have allowed one to measure extremely low concentrations of pharmaceuticals in surface water, rivers, streams, etc. [34]. The occurrence of organic wastewater contaminants is high in the environment, 80% of 139 streams sampled in the USA [35] showed at least one organic wastewater contaminant, although the authors pointed out that the results were influenced by the design of the study and it can not be considered as representative of the global situation in USA streams. The concentrations were, in general, less than 1 μg L^{-1} but their presence in many streams indicated that compounds survived biodegradation.

Pharmaceuticals in effluents of wastewater treatment plants are diluted when entering river waters being detected in the ng L^{-1} range. However, the same spectrum of compounds that are found in the STP are found in the Ebro river basin where analgesics (diclofenac, naproxen, ibuprofen), lipid regulators (gemfibrozil, bezaibrate), antibiotics (azithomycin, trimethoprim, and sulfamethoxazole), antipiletic (carbamezapine), antihistamic (ratidine), and β-blockers (atenolol and sotanol) are the recorded compounds, which are consumed at high levels in Spain [32]. Drugs in a large body of receiving water are in many cases below detection limits although in small receiving streams were around 15–30% effluent median concentration [36]. The availability of occurrence data for pharmaceuticals in estuarine or marine waters is less common than stream and river waters. In the North Sea, for clofibric acid concentrations of $1\,\mathrm{ng\,L^{-1}}$ have been reported, whilst in seawater samples ibuprofen has not been measured above $0.2\,\mathrm{ng\,L^{-1}}$ [37, 38]. Pharmaceutical residues are present as contaminants in UK estuaries [39], but the authors only detected above the detection limits the following targeted compounds/metabolites: clofibric acid, clotrimazolem dextropropoxyphene, dicofenac, ibuprofen, mefenamic acid propanolol, tamoxifen, and trimethoprim, with ibuprofen showing the highest detected concentration ($928\,\mathrm{ng\,L^{-1}}$). In the Victoria Harbor of Hong Kong, antibiotics (belonging to the class quinolones, macrolides, sulfonamides, β-lactam, and chloramphenicol) were mainly below the limit quantification (LOQ). However, they were found in the Pearl River during the high and low water seasons in the range $10–100\,\mathrm{ng\,L^{1}}$. The level of antibiotics in the high water season is controlled by daily sewage discharge patterns and in the low season may be controlled by water column dynamics [40].

There is less knowledge about pharmaceutical concentrations in soil and sediment than for the aquatic environment. This was due to the lack of suitable sensitive analytical methods for the detection of compounds [41]. The persistence of a drug in a sediment or soil mostly depends on its photostability, its binding and adsorption capability, its degradation rate, and leaching in water [42]. The main route of entry for antibiotics for human use is related to the use of sewage sludge for fertilizing the soil. The occurrence of fluoroquinolones, ciprofloxacin, and norfloxacin in sewage sludge has been detected at concentrations ranging between 1.4 to $2.4\,\mathrm{mg\,kg^{-1}}$ [43], which is in the same range as can be measured in digested sludge, indicating a high affinity to the solid phase. Most of the literature on pharmaceuticals in solid environmental samples is related to veterinary drugs, especially those employed in fish farming, which are principally antibiotics.

Pharmaceuticals, as other chemical compounds, can be accumulated by aquatic or benthic organisms. Oxytetracycline (OTC, tetracycline) and oxolonic acid (OA, quinolone) are accumulated by the blue mussel, preferentially being accumulated in the viscera for OTC and in the gills for OA. Bioaccumulation factors (BAF) were low (< 0.5) regardless of the analyzed bivalve part. The application of K_{ow} for antibiotic bioaccumulation can predict a weak accumulation in mussel for antibiotics with $K_{\mathrm{ow}} < 2$, whereas antibiotics such as macrolides with $K_{OW} > 2$ accumulate at a higher level [44]. Fluoxetine and sertraline are prescribed as antidepressants and their occurrence has been detected in surface water or effluent discharges [35, 45]. The analysis of these compounds in streams from a reference site and an effluent-dominated stream showed that these compounds were not detected in the reference site whereas they were detected in all tissues analyzed from fish from the effluent-dominated stream, including *P. nigromaculatus*, L. *macrochirus*, and *I. punctatus*, with a preferential accumulation in the brain, although they also accumulate in muscle at concentrations higher than the limits of quantitation, and subsequently an exposure route to humans in this way should be considered [46]. The influence of pH on the bioconcentration factor of fluoxetine in the fish *Oryzia latipes* has been analyzed [47], showing that BCF values were lower at pH 7 and higher at pH 9 because of an increase of hydrophobicity at pH values closer to pK_a.

4 Ecotoxicology of Human Pharmaceuticals

4.1 Acute Toxicity

Aquatic organisms are targets to analyze the effect of human pharmaceuticals because they are exposed via wastewater over their whole life. Drugs are designed to have a specific mode-of-action along the target pathway. Hypotheses about the mode-of-action in lower animals in many cases are not well supported, because many of the organisms lack the required receptors. Although a mode-of-action for a pharmaceutical should be taken into account when an experiment is designed, this approach may not be appropriate because the mode-of-action could be different or not well known [48].

The ecotoxicological effects of human pharmaceuticals are focused on acute and standard tests. More than three-hundred-and-six endpoints for pharmaceutical ecotoxicity data have been collected for macroinvertebrates, fish, and algae, and over one-hundred for human pharmaceuticals [12]. The selection of three trophic levels (algae, *Daphnia*, and fish) showed that sensitivity followed the order algae > *Daphnia magna* > fish. However, the range of acute toxicity endpoints varied from $> 15\,000\ \mathrm{mg\,L^{-1}}$ (for atropine sulfate-anthicolorgenic/mydriatic) [49] to $< 0.003\ \mathrm{mg\,L^{-1}}$ for fluvoxamine (antidepressant) [50]. The ecotoxicity effects for therapeutic classes showed the following order: antidepressants, antibacterials, and antipsychotics [12]. A recent review [48] summarized the ecotoxicity data, taking into account the ecological relevance and the different classes of human pharmaceuticals: analgesic and non-steroidal anti-inflammatory drugs, beta-blockers, blood lipid-lowering agents, neuroactive compounds, and cytostatic compounds and cancer therapeutics. Seventeen percent showed acute toxicity below 100 mg L^{-1} and 38% above 100 mg L^{-1}, which is classified as not harmful for aquatic organisms according to EU Directive 93/67/EEC. The rest of the compounds (45%) showed high variability in acute toxicity tests. The difference between the acute toxicity data and the environmental levels for human pharmaceuticals demonstrate that only in the case of spills will the toxicity be relevant.

4.2 Chronic Toxicity

The standard acute toxicity tests have as endpoints the lethality and they do not seem appropriate for risk assessment of pharmaceuticals, because of the nature of these compounds. The use of chronic tests over the life-cycle of organisms for different trophic levels could be more appropriate [51]. Nevertheless, the database for this kind of bioassay is very limited.

Most chronic aquatic toxicity data for human pharmaceuticals are available for algae because they are the quickest to perform and therefore less expensive. The sensitivity to antimicrobial substances is higher in Cyanobacteria such as *Microcystis aureginosa* than standard algal toxicity tests (*Pseudokirchneriella subcapitata*) although there are no differences for non-antimicrobial substances [52].

Only in the case of the synthetic steroid EE2, which is present in contraceptive pills, has an effect been observed at environmentally relevant concentrations. In a recent study [53], vitellogenin induction in fathead minnows was reported at an EC_{50} value of 1 ng L^{-1}. The life-cycle exposure of zebrafish to 3 ng L^{-1} EE2 provoked an increase of vitellogenin and caused gonadal feminization [54]. The exposure of some invertebrate taxa (snails) to EE2 also caused effects at very low concentrations $\sim$ 1 ng L^{-1} [55]. Fish are also sensitive to other sex hormones such as methyltestosterone and beta-adrenergic receptor blockers [56].

Analgesic and non-steroidal anti-inflammatory drugs are the most-consumed drugs, and a chronic study with diclofenac has been reported in invertebrates [22, 57]. A chronic study with rainbow trout showed renal lesions at 5 μg L^{-1} [58]. Regarding beta-blockers, propanolol showed chronic toxicity not only on the cardiovascular system in fish but also in the reproductive system [48]. The number of eggs released by fish was reduced at 0.5 μg L^{-1} after four weeks of exposure but not at 50 and 100 μg L^{-1} [59]. The blood lipid-lowering agents have been evaluated by traditional toxicity tests and NOEC in the range of 246 μg L^{-1} to 70 mg L^{-1} have been recorded for *B. caliciflorus* (2 days) and early life stages of zebrafish (10 days), respectively [57].

Chronic toxicity tests have been carried out with carbamezapine (an antiepileptic) and *C. dubia* showed a NOEC (7 days) = 25 μg L^{-1} [57]. Lethal concentration in zebrafish was reported at 43 μg L^{-1} [60]. Chronic studies have been carried out on selective serotonin re-uptake inhibitors (SSRI). Serotonin is a neurotransmitter found in vertebrates and invertebrates. SSRI may affect the function of the nervous and associated hormonal systems. The role of serotonin varies between phyla and in consequence also the effects of SSRI; in medaka (*O. latipes*) serotonin induced oocyte maturation [61] but the opposite action was reported in mummichog (*F. heteroclitus*) [62]. The chronic effects of SSRI on reproduction in fish and invertebrates are not yet clear, interference in the reproduction occurred at concentrations not ecologically relevant [48].

To date, chronic toxicity data using marine or estuarine species have been very scarce. The results with different classes of compounds (carbamezapine, acetaminophen, and ibuprofen) and the endpoint inhibition growth at 72 h for the marine microalgae *Phaeodactylum tricornutum* did not show toxicity below 2.0 mg L^{-1}.

Studies concerning the effects of mixtures of pharmaceuticals are very limited in the scientific literature [63, 64]. The mixture of diclofenac, ibuprofen, naproxen, and acetylsalicylic acid has been evaluated using *Daphnia* and algae, the toxicity of the mixture followed the concept concentration addition. Nevertheless, the effects of mixtures of compounds with different modes-of-action depends on the species and they do not all act in the same way. Few studies concerning the toxicity of mixtures of pharmaceuticals in realistic ecological systems (microcosms and mesocosms) have been carried out. The effect of a combination of eight pharmaceuticals at three levels on *Lemna gibba* and *Myriophyllium sibiricum* has been tested [65]. In a similar microcosm (periphyton, phytoplankton, zooplankton, algae, and benthic communities), three pharmaceuticals with different modes-of-action were analyzed at three levels [66]. At low concentrations (6–10 μg L^{-1}) only trends were appreciable and no significant effects could be recorded. The comparison of assayed treatment with current concentrations in the environment did not allowed to establish a risk situation for this mixture. Nevertheless, many pharmaceuticals are present in the environment and the effect of this "cocktail" could affect to aquatic communities.

5 Environmental Risk Assessment

The objective of environmental risk assessment is to determine the nature and likelihood of the effects of human actions (in this case the use of pharmaceuticals) on animals, plants, and the environment [67]. According to this principle, operational monitoring in support of this concept should be adequate for characterizing exposure and effects [68]. The two-tiered approach (EU and USA) is employed normally for risk assessment of pharmaceuticals (see Sects. 2.1 and 2). In both risk strategies trigger values are selected for further research via tiered assessment 0.01 $\mu g\,L^{-1}$ and 0.1 $\mu g\,L^{-1}$, respectively. The use of this value permits a reduction in the need to carry out many assessments which facilitates the release of new drugs to the market. However, for some compounds this trigger value is insufficient; this is the case for endocrine disruptors which at 1 $ng\,L^{-1}$ showed environmental effects, below the stricter trigger value.

The potential effect of pharmaceuticals is calculated according to the ratio between PEC and PNEC. The PEC is calculated in many cases using figures such as sales, density of population, etc., representing the worst case. In order to get a refinement of this value more precise environmental risk assessment should be carried out; data for biodegradation adsorption, and abiotic factors (pH, temperature) of the environment should be taken into account. The use of measured concentrations allows one to establish more realistic ERA. The other data which should be available is the PNEC, but the lack of chronic toxicity data has made it difficult to perform this assessment. The use of the assessment factor when only acute data are available involves the reduction of uncertainty associated with its use [22]. Though the use of a quantitative structure—activity relationship has been pointed out as a possibility for identifying hazard or prioritizing substances to be analyzed it is not sufficiently precise for risk assessment [48].

The risk of an acute toxic effect from pharmaceuticals in the environment is unlikely [21]. However, many drugs have been designed to affect specific biological systems in target organisms at relatively low dose and exposure concentrations. For this reason, the long-term sublethal effects of pharmaceuticals could be a greater potential concern than acute effects. With the exception of a limited number of drugs, available ecotoxicity data could be inadequate for risk assessment and an extensive suite of chronic sublethal tests may be necessary [69].

6 Concluding Remarks

Although human pharmaceuticals are found at low concentrations in the environment and acute toxicity is not frequent, a broad database with chronic and subtle toxicity tests is necessary to carry out the ERA of these compounds. A priori knowledge about the target biological pathway can identify compounds with higher priority for testing and the species, life stages, and endpoints suitable for testing. In this sense, the selection of estuarine and marine species should be considered.

On the other hand, biomarkers as responses to molecular or biochemical changes can be useful for ecological risk assessment. In vitro systems can be appropriate tools for screening the ecotoxicological effect of pharmaceuticals before fish toxicity testing is carried out. The lack of toxicity tests for pharmaceutical mixtures should be taken into account in order to improve the risk assessment because of the additive, antagonistic, or synergetic effects that can be present. Finally, new technologies such as proteomics and genomics, which are powerful tools for human diagnosis, are under development and they may be helpful to validate effects in the environment and should be included in the framework of ERA, although its use is limited by the current knowledge of the impacted biota.

References

1. Field JA, Johnson CA, Rose JB (2006) Environment Sci Technol 40:1
2. Chapman PM (2006) Environ Toxicol Chem 25:2
3. Daughton CG, Termes TA (1999) Environ Health Perspect 107:907
4. CPMP (Committee for Propietary Medicinal Products) (2003) Note for Guidance on Environmental Risk Assessment of Medicinal Product for Human Use. CPMP/SWP/4447/00. EMEA, London
5. Halling-Sorensen B, Nielsen SN, Lanzky PF, Ingerslev F, Holten Lützhoft HC, Jorgensen SE (1998) Chemosphere 36:357
6. Díaz-Cruz S, Barceló D (2004) Ocurrence and Analysis of Selected Pharmaceuticals and Metabolites as Contaminants Present in Waste Water, Sludge and Sediments. In: Barceló D (ed) The Handbook of Environmental Chemistry, vol 5. Springer, Berlin Heidelberg New York, p 227
7. Ternes T, Hirsch R, Mueller JM, Haberer K (1998) Fresen J Anal Chem 362:329
8. Ternes T, Bonerz M, Schmidt T (2001) J Chromatogr A 938:175
9. Kolpin DW, Furlong ET, Meyer MT, Thurman EM, Zaugg SD, barber LB, Buxton HT (2002) Env Sci Technol 36:1202
10. Fent K, Weston AA, Caminada D (2006) Aquat Toxicol 76:122
11. Cunningham VL, Constable DJC, Hannah RE (2004) Environ Sci Technol 38:3351
12. Webb SF (2004) A data based perspective on the enviromental risk assessment of human pharmaceuticals I—Collation of available ecotoxicity data. In: Kümmerer K (ed) Pharmaceuticals in the Environment, Springer, Berlin, p 317
13. Crane M, Watts C, Boucard T (2006) Sci Total Environ 367:23
14. EMEA (European Medicines Angency) (2006) Guideline on the Environmental Risk Assessment of Medicinal Products for Human Use. Doc Ref EMEA/CHMP/SWP/4447/00. EMEA, London
15. US Environment Protection Agency (1998) Guidelines for Ecological Risk Assessment EPA/630/R-95/002F. USEPA, Washington DC
16. OECD (1996) Guidelines for testing chemicals. OCDE Publications Service, Paris, p 1040
17. EMEA (2007) European Risk Management Strategy: 2008–2009 work programme adopted. Doc Ref EMEA/564868/2007. EMEA, London
18. Guidance for Industry Environmental Assessment of Human Drug and Biologics Application (1998) http:// www.fda.gov/cder/guidance/index.htm, Accessed 25 March 2008
19. Webb SF (2004) A data based perspective on the enviromental risk assessment of human pharmaceuticals II—Aquatic Risk Characterisation. In: Kümmerer K (ed) Pharmaceuticals in the Environment, Springer, Berlin, p 345

20. Huschek G, Hansen PD, Maurer H, Krengel D, Kayser A (2004) Environ Toxicol 19:226
21. Carlsson C, Johansson AK, Alvan G, Bergman K, Kühler T (2006) Sci Total Environ 364:67
22. Ferrari B, Mons R, Vollar B, Frayse B, Paxeus N, Giudice RL, Garric J (2004) Environ Toxicol Chem 23:1344
23. Jones OAH, Voulvoulis N, Lester JN (2002) Water Res 36:5013
24. Dollery CT (1991) Therapeutic drugs, vol I and II. Churchill Livingstone, Edinburgh
25. Richardson ML, Bowon JM (1985) J Pharm Pharmacol 37:1
26. Ternes TA, Meisnhiemer M, McDowell D, Sacher F, Braunch HJ, Haist-Gulde B, Preuss G, Wilme U, Zulei-Seilbert N (2002) Environ Sci Technol 36:3855
27. Kot-Wasik A, Debska J, Namiesnik J (2007) Trends Anal Chem 26:557
28. Barcelo D. Trends Anal Chem 26:454
29. Santos JL, Aparicio I, Alonso E (2007) Env Int 33:596
30. Heberer T, Fhurmann B, Schmidt-Bäumler K, Tsipi D, Koutsouba V, Hiskia A (2001) Ocurrence of pharmaceuticals residues in sewage, river, ground and drinking water in Greece and Germany. In: Daughton CG, Jones-Leo T (eds) Pharmaceutical and Personal Care Products in the Environment: Scientific and Regulatory Issues. Symposium Series 791. American Chemical Society, Washington, DC, p 70
31. Farré M, Ferrer I, Ginebreda A, Figueras M, Olivella L, Tirapu M, Vilanova M, Barceló D (2001) J Chromatogr A 938:187
32. Gros M, Petrovic M, Barcelo D (2007) Environ Toxicol Chem 26:1553
33. Jones OAH, Voulvoulis N, Lester JN (2003) Bull WHO 81:768
34. Sedlak DL, Gray JL, Pinkston KE (2000) Environ Sci Technol 34:509A
35. Kolpin DW, Furlng ET, Meyer MT, Thurman EM, Zaug SD, Barber LB, Buxton HT (2002) Eviron Sci Technol 3:1202
36. Brun GL, Bernier M, Losier R, Doe K, Jackman P, Lee HB (2006) Environ Toxicol Chem 25:2163
37. Buser HR, Poiger T, Müller MD (1999) Environ Sci Technol 33:2259
38. Weigel S, Kulhmann J, Hühnrfuss H (2002) Sci Total Environ 295:131
39. Thomas KV, Hilton MJ (2004) Mar Pollut Bull 49:436
40. Xu W, Zhang G, Zou S, Li X, Liu (2007) Environ Pollut 145:672
41. Hamscher G, Pawelzik HT, Höper H, Nau H (2004). Antibiotics in soil: routes of entry, environmental concentration, fate and possible effects. In: Kümmerer K (ed) Pharmaceuticals in the Environment. Springer, Berlin, p 139
42. Díaz Cruz MS, López de Alda MJ, Barceló D (2003) Trends Anal Chem 22:340
43. Golet EM, Sthehler A, Alder AC, Giger W (2002) Anal Chem 74:5455
44. Bris HL, Puliquen H (2004) M Pollut Bull 48:434
45. Metcalfe CD, Miao XS, Koening BG, Struger J (2003) Environ Toxicol Chem 22:2281
46. Brooks BW, Chamblis CK, Stanley JK, Ramirez A, Banks KE Johnson RD, Lews RJ (2005) Environ Toxicol Chem 24:464
47. Nakamura Y Yamamoto H, Sekizawa J, Kond T, Hiai N, Tatarazako N (2008) Chemsphere 70:865
48. Fent K, Weston AA, Caminada D (2006) Aquat Toxicol 76:122
49. Calleja MC, Persoone G, Geladi P (1994) Arch Environ Contam Toxicol 26:69
50. Fong PP, Huminski PT, D'Urso ML (1998) J ExpZool 280:260
51. Halling-Sørensen B; Nielsen N, Lanzky PF, Ingerslev F, Holten-Lützhøft HC, Jørgensen SE (1998) Chemosphere 36:357
52. Boxakk ABA, Kolpin DW, Halling-Sørensen B, Tolls J (2003) Environ Sci Technol 37:286

53. Brian JV, Harris CA, Scholze M, Backhaus T, Booy P, Lamoree M, Pojana G, Jonkers N, Runnalls T, Bonfa A, Marcomini A, Sumpter JP (2005) Environ Health Perspect 113:721
54. Fenske M, Maack G, Schafers C, Segner H (2005) Environ Toxicol Chem 24:1088
55. Schulte-Oehlmann U, Oekten M, Bauchmann J, Oehlmann J (2004) Effects of ethinyloestradiol and methyltetosterone in prosobranch snails. In: Kümmerer K (ed) Pharmaceuticals in the Environment. Springer, Berlin, p 237
56. Zerulla M, Länge R, Steger-Hartmann T, Panter G, Hutchinson T, Dietrich DR (2002) Toxicol Lett 131:51
57. Ferrari B, Paxeus N, Lo Giudice R, Pollio A, Garric J (2003) Ecotoxicol Environ Safe 55:359
58. Schawaiger J, Ferling H, Mallow U, Wintermayr H, Negale RD (2004) Aquat Toxicol 68:141
59. Huggett DB, Brooks BW, Petreson B, Foran CM, Sclenk D (2002) Arch Environ Contam 43:229
60. Thaker PD (2005) Environ Sci Technol 39:193A
61. Iwamatsu T, Toya Y, Sakai N, Terada Y, Nagata R, Nahahama Y (1993) Dev Growth Differ 35:625
62. Cerda J, Subhedar N, Reich G, Wallace RA, Selman K (1998) Biol Reprod 59:53
63. Cleuvers M (2003) Toxicol Lett 142:185
64. Cleuvers M (2004) Ecotoxicol Environ Safe 59:309
65. Brain RA, Johnson DJ, Richards SM, Hanson ML, Sanderson H, Lam MW, Young C, Mabury SA, Sibley PK, Solomon KR (2004) Aquat Toxicol 70:23
66. Richards SM, Wilson CJ, Johnson DJ, Castle DM, Lam M, Mabury SA, Sibley PK, Solomon RK (2004) Environ Toxicol Chem 23:1035
67. Jones OA, Volvoulis N, Lester J (2004) Cri Rev Toxicol 47:71
68. Hansen PD (2007) Trends Anal Chem 26:1095
69. Ankley GT, Brooks BW, Huggett DB, Sumpter JP (2007) Environ Sci Technol

Subject Index

MIX
Papier aus verantwortungsvollen Quellen
Paper from responsible sources
FSC® C105338

If you have any concerns about our products,
you can contact us on
ProductSafety@springernature.com

In case Publisher is established outside the EU,
the EU authorized representative is:
Springer Nature Customer Service Center GmbH
Europaplatz 3, 69115 Heidelberg, Germany

Printed by Libri Plureos GmbH
in Hamburg, Germany